CHAPTER I.
COMPOSITION AND GENERAL STRUCTURE OF THE BODY.

Anatomy is the study of the physical structure and physiology the study of the normal functions of the human body.

Chemical Constitution of the Body.—In the body only twenty elements have been found. These include carbon, oxygen, hydrogen, nitrogen, sulphur, phosphorus, calcium, magnesium, manganese, chlorin, potassium, and fluorin. For the most part they appear in very complex and highly unstable combinations, though oxygen and nitrogen may be said to exist uncombined in the blood, alimentary canal, and lungs. Hydrogen also occurs in simple form in the alimentary canal, but as the result of fermentation, not as an element of the body.

Of the organic compounds some contain nitrogen and some do not. The most important of the former are the proteins, which are found only in living bodies and consist of carbon, hydrogen, oxygen, nitrogen, and sulphur combined in very similar proportions. The important proteins in the body are the serum albumen and fibrin found in the blood, myosin in muscle, globulin in the red blood-corpuscles, and casein in the milk. Similar to the proteins but capable of passing through membranes are the peptones, the final result of protein digestion, from which the albuminoids differ in that they contain no sulphur. Ferments containing nitrogen exist in all the cells of the body, though more particularly in those of the digestive organs, and the coloring matters, as the bilirubin of the bile, are nitrogenous.

The organic substances that do not contain nitrogen are the carbohydrates or starches, the hydrocarbons or fats, and the acids, of which the most important is carbon dioxide, given off by the lungs.

The inorganic substances are water, which forms a large percentage of all the tissues and from one-fourth to one-third of the whole body weight, sodium chloride or common salt, which plays an important part in keeping substances in solution, potassium and magnesium chloride, and hydrochloric acid, found in the stomach.

The Cell.—Although the body is a very complex organism, the cell is its unit or foundation. In fact, the body begins life as a single protoplasmic cell, the ovum, which is frequently compared to the amœba, a microscopic animal consisting of a single cell of protoplasm or living substance—a substance not well understood as yet—but possessing practically all the functions of the human body. For, although it has no organs and is homogeneous in structure, the amœba can move by throwing out a process, and can surround and absorb food, which it builds up into new tissue, discarding the waste. The ovum, however, differs from the amœba in that it has a transparent limiting membrane and contains a darker spot, the nucleus. This in turn contains another smaller spot, the nucleolus, while through the protoplasm, which is semi-fluid, extends a fine network that seems to hold it in place.

The *ovum* is very small, about $1/125$ inch in diameter, and after fertilization grows by segmentation, the nucleus dividing in two and the protoplasm grouping itself anew about the two nuclei. This division continues, each cell dividing and forming two, or sometimes four, new cells, all of which at first appear alike. By degrees, however, differentiation takes place and different groups of cells assume different characteristics. Thus the various tissues are gradually developed, each with a structure and a function of its own, and are distributed among the various organs, each organ consisting of several tissues. During the process of growth and even after full growth of the body is attained old cells are continually dying and being replaced by new ones.

The typical cell is circular, but through being squeezed together in the tissues or for some other reason the cells vary in shape in different parts, being at times hexagonal, spindle-shaped, or columnar. Yet, whatever their differences in shape or other characteristics, they all live the same sort of life. All protoplasm absorbs oxygen when it comes in contact with it and in the process of combining with it is in part burned or oxidized, with the consequent setting free of heat and other forms of energy and the formation of carbon dioxide. So long as the body is alive, therefore, whether it is in a state of activity or of rest, it is the seat of constant chemical change throughout all its cells, and to these chemical changes are due all the forms of energy manifested by the body. For energy is never destroyed, though it may appear in a different form, and the elements of the human body are so combined that their energy may be liberated and manifested in the different functions the body exhibits.

The fundamental tissues of the body are the epithelial tissues, the connective tissues, including the cartilaginous and bony tissues, and the muscular and nervous tissues. Of these the epithelial tissues serve as a protection to the surface of other tissues; the connective tissues together form a framework for the support and general protection of the other tissues; while energy is expended by muscular and nervous tissue, the latter directing the former in its movements. All the tissues are inter-dependent and the organs work together. Besides cells every tissue contains a certain amount of lifeless matter, the intercellular substance, which was at some time produced by the cells.

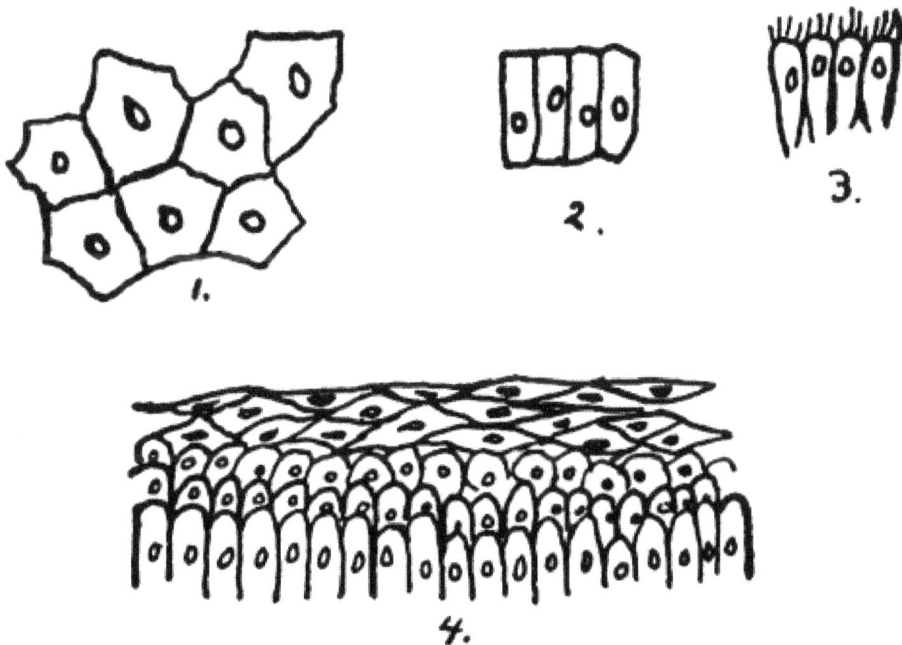

FIG. 1.—Epithelium: 1, pavement epithelium; 2, columnar epithelium; 3, ciliated epithelium; 4, stratified epithelium.

In **epithelial tissue** there is little intercellular substance, the cells being close together and arranged generally as a skin or membrane covering external or internal surfaces. When there are several layers of cells, the deepest are columnar in shape and the others become more and more flattened and scale-like as they approach the surface, where they are gradually rubbed off and replaced by the growth of new cells from below. This stratified epithelium, as it is called, is found wherever a surface is exposed to friction, as in the skin and in the mucous membrane of the mouth, pharynx, and esophagus, and in that of the vagina and the neck of the uterus. In simple epithelium, where there is only a single layer of cells, the cells may be pavement or hexagonal, columnar, glandular, or ciliated, according to their different functions. The flat pavement cells occur where a very smooth surface is required, as in the heart, lungs, blood-vessels, serous cavities, etc. None of these surfaces communicate directly with the external surface of the body and the name endothelium is substituted for epithelium. The columnar form of cell in the intestine facilitates the passage of leucocytes between the cells. In glandular epithelium the cells vary according to the gland in which they occur, their protoplasm being filled with the material the gland secretes. Finally, ciliated epithelium is composed of columnar cells with cilia or little hair-like processes upon their free surface which serve to send secreted fluids and other matters along the surfaces where they occur, as in the air passages, parts of the generative organs, the ventricles of the brain, and the central canal of the spinal cord.

Connectivetissue has a great deal of intercellular substance. One form, *areolar tissue,* is composed of a loose network of fine white fibers with a few yellow elastic fibers interspersed and with cells lying in the spaces between the fibers. It connects and surrounds the different organs and parts, holding them together, yet allowing free motion, and is one of the most extensively distributed of the tissues. It is continuous throughout.

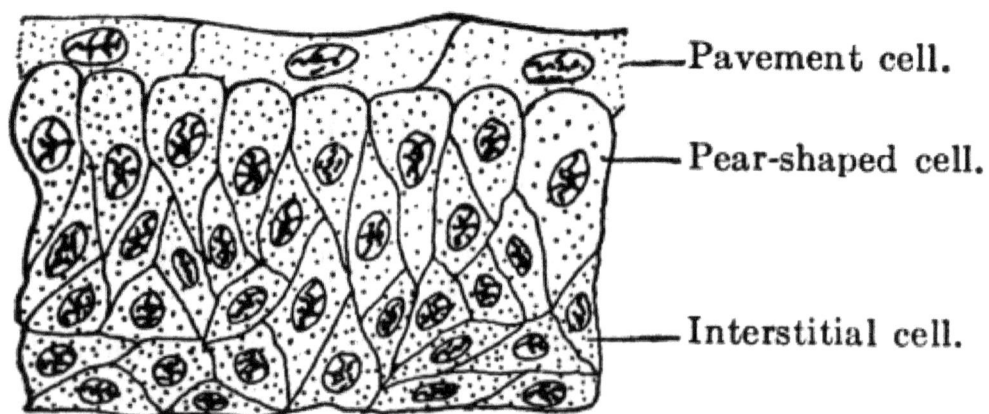

FIG. 2.—Section of bladder epithelium. (Hill.)

Closely allied to the areolar is the _fibrous tissue_, in which the white fibers lie close together and run for the most part in one direction only. This is found in ligaments, joints and tendons, as also in such fibrous protective membranes as the periosteum, dura mater, the fasciæ of muscles, etc. Fibrous tissue is silvery white in appearance and is very strong and tough, yet pliant. It is not extensile.

Elastic tissue, on the other hand, has a large predominance of yellow elastic fibers and is very extensile and elastic, though not so strong as the fibrous. It is found in the walls of the blood-vessels, especially the arteries, in the walls of the air tubes, in the ligaments of the spine, etc.

Fatty or adipose tissue is formed by the deposit of fat in the cells of the areolar tissue and is found in most parts where the areolar tissue occurs, though it varies largely in amount in different parts. It is found pretty generally under the skin, fills in inequalities about various organs and about the joints, and exists in large quantities in the marrow of the long bones. In moderate amounts it gives grace to the form and constitutes an important reserve fund.

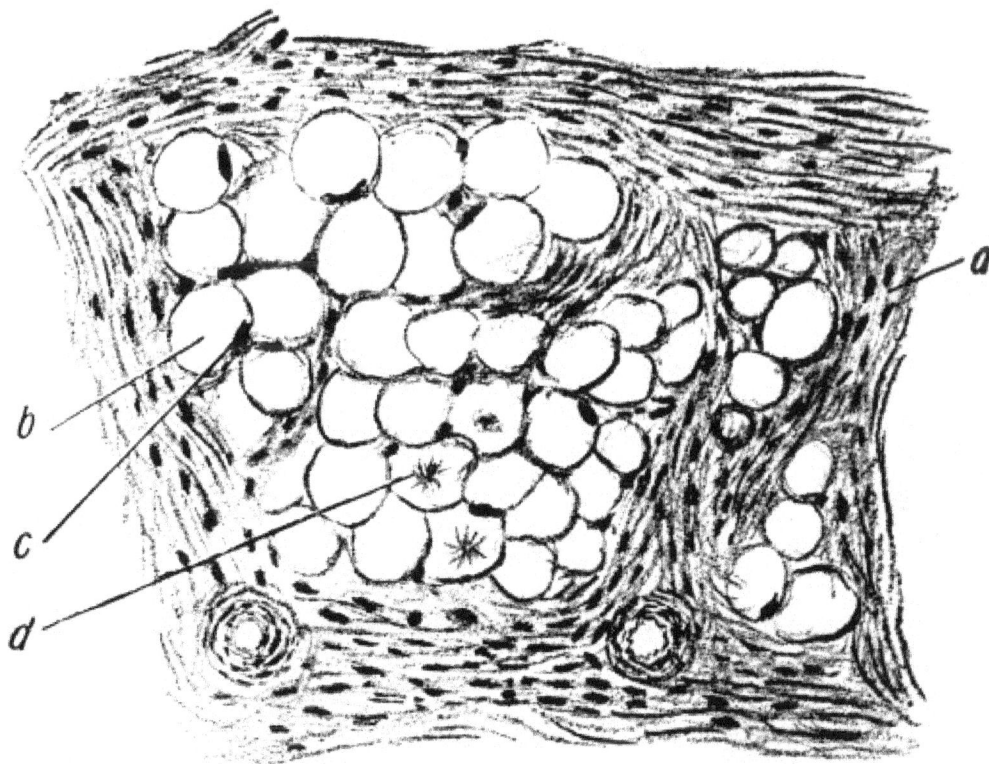

FIG. 3.—Adipose tissue (Leroy): _a_, Fibrous tissue; _b_, fat cells; _c_, nucleus of fat cells; _d_, fatty acid crystals in fat cells.

Cartilage consists of groups of nucleated cells in intercellular substance. It is very firm, yet highly elastic, and serves in the joints to break the force of concussion of the harder and less elastic bones. Except when it occurs at the end of a bone, it is covered with a membrane called the perichondrium, which carries its blood supply. In the nose, ear, larynx and trachea it serves to give shape, to keep the passages open, and to afford attachment for muscles. Most of the skeleton of the fetus consists of cartilage, which later develops into bone.

Bone.—In bone the intercellular tissue is rendered hard by the deposit of mineral salts, the resulting material being of great strength and rigidity. The texture may be close and dense like ivory or open and spongy, the difference lying merely in the fact that the one has fewer spaces between the solid particles than the other. There is usually a hard, compact layer on the exterior of the bone, as that is where the greatest cross-strain comes, especially in the long bones, while within is the cancellous or spongy tissue, which gives lightness to the bone and is capable of withstanding enormous pressure, though it can bear little cross-strain.

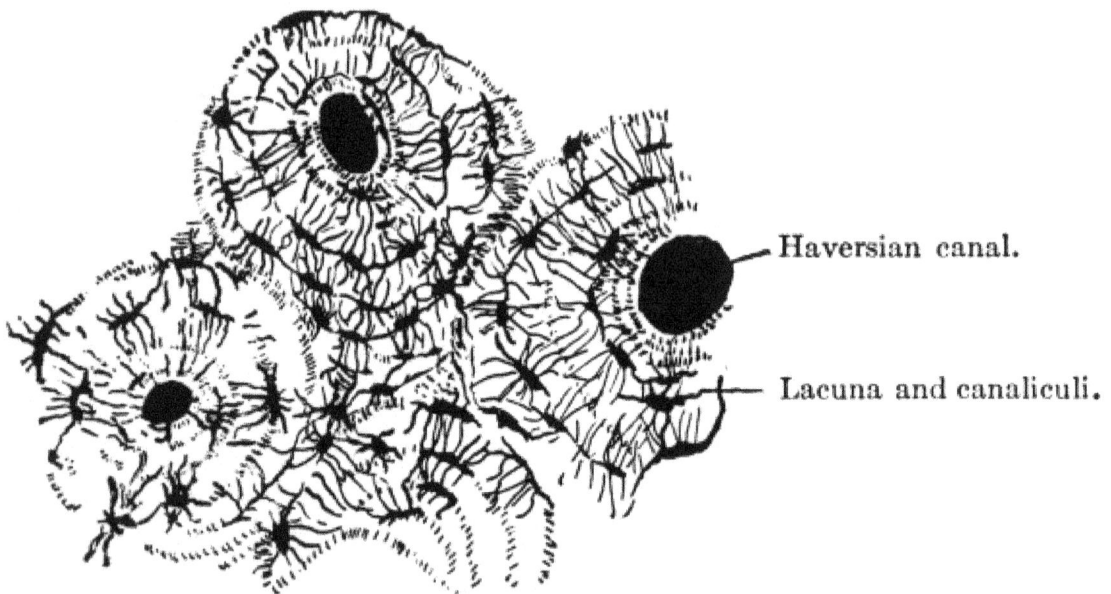

— Haversian canal.

— Lacuna and canaliculi.

FIG. 4.—Cross-section of compact bone tissue. (After Sharpey.)

Structure of Bone.—The hard substance in bone is always arranged in lamellæ or bundles of bony fibers, which in cancellous tissue meet to form a kind of lattice-work, while in the dense tissue they are generally arranged in rings about the Haversian canals, channels through which the blood-vessels pass through the bone longitudinally. Between the lamellæ are spaces called _lacunæ_, in which lie branched cells, the spaces being connected with each other and with the Haversian canals by numerous tiny canals or canaliculi, by which nutrient material finds its way from the Haversian canals to all parts of the bone.

Within the bone is the medulla or _marrow_, which is of two varieties: the yellow, which is largely fat and is found in the long bones of adults, and the red, which is nearly three-fourths water and is found in most of the other adult bones and in the bones of the fetus and of the infant.

Lining the medullary and cancellous cavities is a delicate connective tissue lining, the _endosteum_, which contains many bone-forming cells, while on the outside of the bone, except at the articular ends, is the _periosteum_ with its outer protective layer and its inner vascular layer containing osteoblasts or bone-forming cells. The periosteum is essential for the growth of new bone where the old bone has died, and if the periosteum is removed from healthy bone the part beneath is liable to die, as it is by the constant growth of the osteoblasts that the bone grows and is renewed. In the repair of broken bones tissue is formed between and around the broken ends.

Bone Formation.—Most of the skull and face bones begin as membranes of connective tissue, that is, are formed in membrane. Bones are also formed in cartilage, the bone formation in this case beginning from centers of ossification, where the deposit of lime salts in the intercellular substance begins, the salts coming to the centers dissolved in the plasma. Such a center of growth in a bone is called the epiphysis and is separated from the main part of the bone or diaphysis by cartilage until full growth is attained, when ossification becomes complete. So in surgery, in working on the bones of children, part of the epiphysis should always be left for the sake of future growth. The outer shell of compact tissue is deposited by the periosteum.

Chemical Composition of Bone.—Chemically bone is composed of about one-third organic or animal matter, largely gelatine, and two-thirds inorganic matter, including various salts of calcium, magnesium, and sodium. In young children the animal matter predominates and the bones are soft and often bend instead of breaking, only the outside shell on one side giving way, as in "green-stick" fracture. In rickets there is a deficiency of lime salts, but the increased brittleness of the bones in old age is due, not to increase of mineral matter, but to the less spongy texture of old bones.

Classification and Function of Bones.—There are in the body some two hundred bones, which may be classified as long, short, flat, and irregular. Occasionally an irregular bone develops in a fontanelle, the membranous opening at the juncture of the sutures of the skull. This is known as a Wormian bone. It is not, however, included in the two hundred, as are not the sesamoid bones or bones developed in tendons, with the exception of the patella or knee-cap.

Long bones are developed in cartilage and consist of a shaft, two extremities, and various processes. They are more or less curved to give them strength and grace. They serve as supports and act as levers for purposes of motion and the exercise of power. Since a hollow cylinder is just as strong as a solid one of the same size, the weight coming only on the outer shell, the great bones which are accountable for weight and which need to be light themselves have hollow shafts, composed chiefly of compact tissue with a central medullary canal. The ends, however, are expanded in order to make better connection at the joints and to afford broad surfaces for muscular attachment, cancellous tissue being used in them for lightness and strength. The large spongy ends also give elasticity and lessen jar, and by bringing the tendons to the bone at a greater angle increase their effectiveness. Blood is brought to the long bones not only by the vessels of the periosteum but by the medullary artery, which penetrates the compact tissue by the nutrient foramen and divides into an ascending and a descending branch.

Short bones are spongy throughout. They are used for strength and where little motion is required.

Flat bones are composed of two thin layers of compact tissue with a varying amount of cancellous tissue between, and are for protection and muscular attachment. The cancellous material between the two layers or tablets of the skull is called the diploë.

Eminences and depressions occur on bones and when they are not articular are for the attachment of ligaments and muscles. If they are articular, they help to form joints.

As a whole the bony framework serves to keep the soft parts in place, to support and protect them, and to aid in locomotion. The bones of the head and trunk support and protect organs; those of the arms are for tact and prehension; those of the lower extremities are for support and locomotion.

Normally bones have little sensibility, but when inflamed they are extremely sensitive and painful.

Joints.—The bones are connected with and move upon one another by means of joints. These joints are of three kinds: 1. Immovable, where the adjacent margins of the bones are closely applied, with little fibrous tissue between, as in the sutures of the head; 2. those with limited motion, which are very strong, the parts being connected with tough fibro-cartilage; and 3. freely movable. In this last group the articulating surfaces are covered with cartilage, which again is lined with a delicate synovial membrane which secretes a small amount of lubricating fluid, the synovial fluid, to reduce friction. Their surfaces are also sometimes deepened by the presence of inter-articular fibro-cartilages. Bursæ or sacs of synovial membrane occur outside the joints under tendons and ligaments to reduce friction.

The nature and extent of the motion of a joint is defined and the bones are held together by strong bands of fibrous tissue or ligaments, these ligaments being more fully developed in joints where there is great freedom of motion or where there is great weight to be supported. In a ball-and-socket joint, such as the hip, there is a ligament in the form of a strong capsule which surrounds the joint on all sides and limits its motion, while hinge joints, like the elbow, and pivot joints, such as that formed by the atlas on the axis, have lateral ligaments that allow of freer motion. In the shoulder-joint, which is the most freely movable joint in the body, the capsular ligament is very lax.

In general the kinds of motion possible in joints may be said to be flexion, extension, abduction, adduction, circumduction, and rotation.

When much violence is applied to a joint and no dislocation results, as in a sprain, there is often much stretching and even laceration of the ligaments.

Muscle.—The flesh, which forms a large proportion of the weight of the body, consists of muscular tissue. Of this two kinds are found: 1. The striated or striped muscle of animal life, which is under the control of the will and so is known as voluntary muscle, and 2. the unstriped or smooth muscle of organic life over which we have no control, that is, the involuntary muscle. Each fiber of striped muscle has an elastic, membranous sheath, the sarcolemma, and consists of rod-shaped cells with a nucleus along the edge, set end to end and having crosswise striations. In unstriated muscle the fibers, which have no sarcolemma, consist of oval or spindle-shaped cells, with a nucleus much smaller than that of striped muscle and situated in the middle. In both kinds of muscle the fibers are bound together with connective tissue and blood-vessels into fasciculi or bundles, and many bundles go to make up a muscle. The muscle in turn has a connective tissue envelope or sheath, the fascia. These fasciæ are found throughout the body, the superficial ones being just beneath the skin, while the deep ones not only form sheaths for the various muscles but form partitions between them and serve to strengthen their attachments. The striped muscles are those of motion, while the unstriped occur in the hollow organs, surrounding the cavity and in some cases lessening its capacity by their contraction.

An intermediate form of muscle known as cardiac muscle occurs in the heart. Here the fibers have striations but the nucleus is generally in the middle of the cell and the fibers branch and run together.

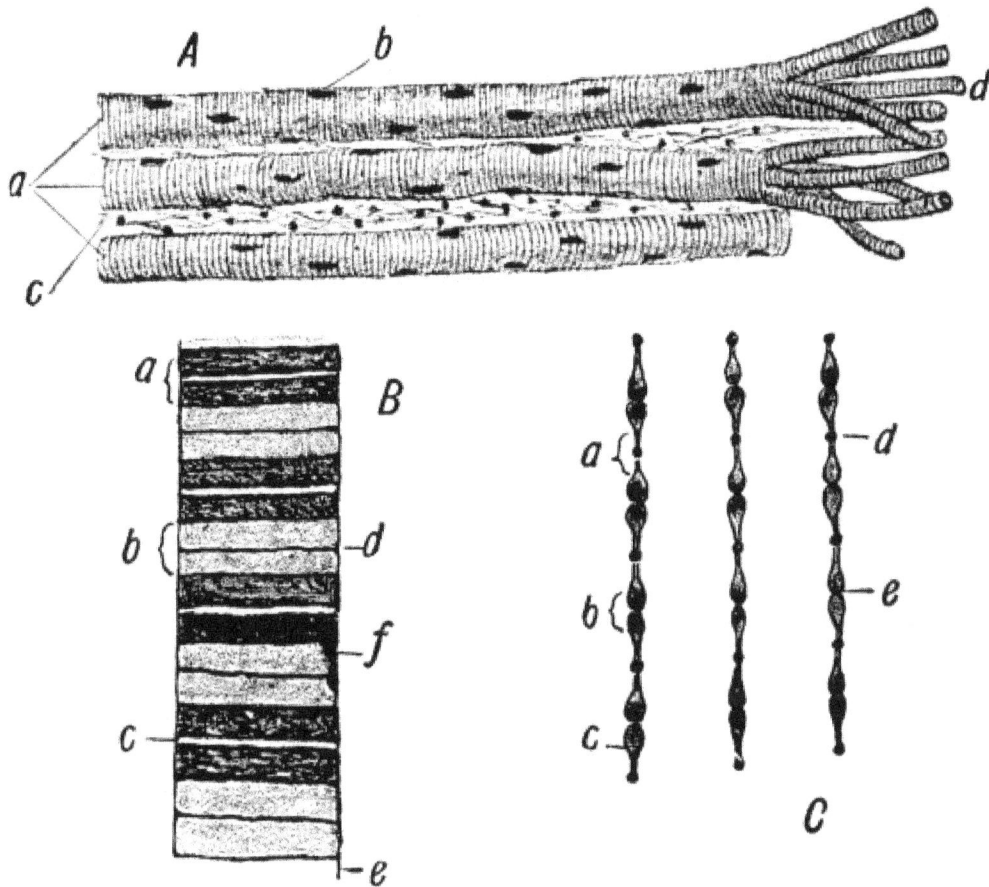

Fig. 5.—Voluntary muscle (Leroy). *A*, Three voluntary fibers in long sections: *a*, three voluntary muscle fibers; *b*, nuclei of same; *c*, fibrous tissue between the fibers (endomysium); *d*, fibers separated into sarcostyles. *B*, Fiber (diagrammatic): *a*, dark band; *b*, light band; *c*, median line of Hensen; *d*, membrane of Krause; *e*, sarcolemma; *f*, nucleus. *C*: *a*, Light band; *b*, dark band; *c*, contracting elements; *d*, row of dots composing the membrane of Krause; *e*, slight narrowing of contracting element aiding in production of median line of Hensen.

In life muscle appears more or less translucent and is contractile and alkaline, but in death it loses its translucency and becomes rigid, at the same time giving off in decomposition much carbon dioxide, so that its reaction is acid. This phenomenon of the muscles becoming rigid in death is called rigor mortis and occurs generally a few hours after death, though it may come at once or be considerably delayed. It may last anywhere from a few moments to several days but generally lasts from twenty-four to thirty-six hours. It is probably due to the formation in the muscle of myosin, a substance which probably comes from myosinogen in the living muscle and which is closely akin to the fibrin of blood. Probably the myosin or what precedes it causes clotting of the muscle just as fibrin or what precedes it causes clotting of the blood.

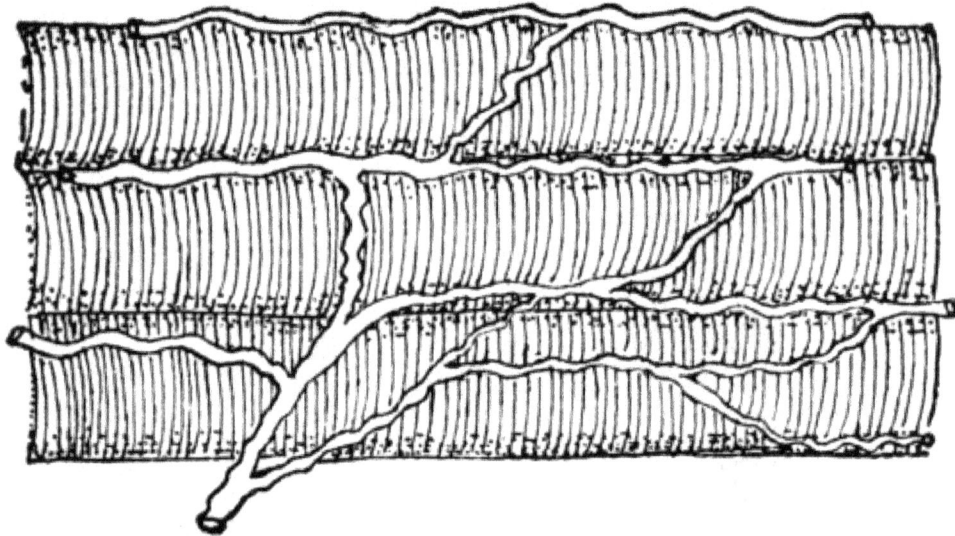

FIG. 6.—Three voluntary muscle fibers from an injected muscle, showing network of blood capillaries. (Hill.)

The muscles vary in shape in different parts of the body, being long and slender in the limbs and broad and flat in the trunk. They are attached chiefly to bones but also to cartilages, ligaments, and skin, either by means of tendons, which are cords or bands of white inelastic fibrous tissue, or by means of aponeuroses, membranous expansions of the same nature. Most voluntary muscles consist of a belly and two ends or tendons. The origin is the fixed point from which it acts while the movable point upon which it acts is known as its insertion.

Action of the Muscles.—When attached to bones, muscles are distributed in three ways: 1. When it is necessary to produce much motion rapidly, a short muscle is used. 2. When a part needs to be moved far and much contraction on the part of the muscle is, therefore, needed, the muscle is very long, as in the case of the sartorius muscle, which shortens half its length. 3. Finally, where less distance has to be covered but greater power is required, tendons are used, as in this case the contraction is powerful but does not carry the part far.

In performing the mechanical work of the body the muscles are aided by the fact that the bones, to which they are largely attached, are set together loosely and form a set of levers, on which the muscles act to perform certain definite acts. All three classes of levers occur: 1. where the fulcrum is between the weight and the power, as in the case of the head, which is balanced by the muscles of the neck on the vertebræ; 2. where the weight is between the fulcrum and the power, as when a person raises himself upon his toes; and 3. where the power is between the fulcrum and the weight, as when the biceps is used to raise a weight held in the hand. The erect position of the body is difficult to maintain because the center of gravity is high up, and it is by the contraction of many muscles in the legs, thighs, back, abdomen, and neck that the body is balanced upright upon the feet.

Physiology of Muscle.—*Irritability* or sensitiveness to stimulation and *contractility* or the power to contract are the two most important functions of muscle. Contraction occurs in response to nervous energy brought by the nerves, a nerve filament going to each muscle fiber, into which it plunges, its substance being lost and its sheath becoming continuous with that of the muscle fiber. Any irritant, as heat, electricity, etc., when applied to the nerve, causes the muscle to contract. Moreover, muscle has an irritability of its own and can contract independently of the nervous system. In contracting it shortens and thickens, bringing the two ends closer together, and becomes firm and rigid. The amount of contraction depends upon the strength of the stimulus and the irritability of the muscle. The minimal stimulus is the least stimulus that will cause a contraction and the maximal is one that will cause the greatest contraction. The work done depends in like manner upon the strength of the stimulus. During contraction certain sounds are given off called muscle sounds, which can be heard with the stethoscope but have no special significance.

The muscles which have the greatest power of rapid contraction are generally attached to levers. Indeed, striated muscle is characterized by the rapidity and strength with which it works, though its rhythmic motion is slight. Smooth muscle, on the other hand, is characterized by its great force, considerable rhythm, considerable tone, and slight rapidity, that is, its contraction is slower and lasts longer than that of striated muscle. Cardiac muscle is characterized by great rhythm and force, fair rapidity, and slight tonicity, tonicity being the amount of tone or readiness to work. For even in sleep muscle is always in tone, that is, ready to do its work. It is this that makes the difference in appearance between a living and a dead person and enables one to spring to his feet at night if he hears a noise, a thing he could not do if his muscles were wholly relaxed. Thus, rapidity is the great function of striated, tonicity of smooth, and rhythm of cardiac muscle. In paralysis the muscles droop and lose their tone. Muscles are frequently the seat of rheumatic disorders.

When set free, potential energy accomplishes work. In muscle there is a good deal of potential energy, which is set free as heat and as work accomplished. Even when the muscles are at rest, chemical changes are going on and heat is being produced, though more heat is produced when they are functioning. If the body depended upon its gross motions for all its heat it would grow cold while a person rested. The respiratory organs, however, and the heart are always working and chemical changes are constantly taking place.

Ordinarily a muscle has some object in contracting, such as the raising of a load, and it contracts voluntarily more or less according to the weight of the load. The amount of work done is calculated in foot-pounds or gram-meters, that is, the energy required to raise one pound one foot or one gram one meter. As a rule the muscles with the longest fibers, as the biceps, do the most work and those with a large number of fibers do more than those with less. It has been calculated that whereas an engine gives back one-twelfth of the energy of the coal consumed, muscle liberates one-fourth of the energy brought to it in the form of food. During activity the glycogen or sugar in the muscle is used up and the muscle becomes more acid, owing to the lactic acid that is formed. The carbon is taken in and carbon dioxide given off. Nitrogen puts the muscle in condition to do its work but is not so much used up in the work as is the carbohydrate material. So it is the non-nitrogenous matter that does the work and any increase in urea, the end-product of protein metabolism, is mere wear and tear.

Sudden heat or cold causes muscular contraction and moderate heat favors both muscular and nervous irritability. Moderate cold, however, lessens the force of contraction and below zero muscle very largely loses its irritability without necessarily becoming rigid.

While well supplied with blood, muscle will contract without fatigue, but if the blood supply is shut off, it soon loses its irritability and becomes rigid. The more a muscle is used in moderation the more it develops, but after it has done a certain amount of work it becomes exhausted, losing its irritability or power to respond to stimuli and later becoming rigid. Such fatigue is due to the production of certain poisonous waste products which have a paralyzing effect on the nerves and which are ordinarily gradually carried away in the blood, but which sometimes, if produced to excess, accumulate too fast for the blood wholly to remove them. Usually the nerve becomes exhausted first and the muscle substance later. So long as it is connected with the nervous system a muscle will respond to stimuli, but when the nerve becomes tired, degeneration is more rapid. In fact, the degree of exhaustion is determined by several factors, as by relation to the central nervous system, variations in temperature, blood supply, and functional activity, the process being more rapid in warm than in cold blooded animals.

Cilia.—A few motions are accomplished by tissue that is not muscular, as in the case of the cilia attached to the cells of the respiratory tract, which lie flat on the free surface and then lash forward, serving in the air cells to keep the air in motion and in the tubes to send secretions from below upward and outward and to keep out foreign bodies. Cilia are also found in the female genital tract, where they aid the passage of the ovum from the ovary to the womb. They act together, though apparently not governed by the nervous system. As in the white corpuscles of the blood, whose motion also is not muscular, the changes that take place in ciliated epithelium are probably about the same as those in muscular tissue, that is, contractile.

The Blood.—To most of the tissues just described nourishment is brought in the blood, which circulates through the body in a system of hollow tubes, the arteries and veins, whence it is distributed through the agency of the lymphatic system. There are no blood-vessels, however, in the epidermis, epithelium, nails, hair, teeth, nor in the cornea of the eye. The vessels that carry the blood from the heart are called arteries, those that return it veins. The former begin as large vessels and gradually decrease in size; the latter begin as small vessels and form larger and larger trunks as they approach the heart.

The *arteries* have three coats: 1. a thin, serous coat, the internal or intima; 2. a middle or muscular coat, and 3. an external coat of connective tissue. The middle coat is the thickest and is the one that prevents the walls from collapsing when cut across. Except in the cranium, each artery is enclosed in a sheath with its vein or veins, the venæ comites. Usually the arteries occupy protected situations and are straight in their course. Where a vessel has to accommodate itself to the movements of a part, however, it may be curved, as in the case of the facial artery which is curled on itself to allow for movements of the jaw. They anastomose or communicate freely with one another, thus promoting equality of distribution and pressure and making good circulation possible even after the obliteration of a large vessel.

The *veins* have three coats like the arteries, but they are not so thick and the muscular coat is not so highly developed, so that the walls collapse when cut and have no elasticity. There are constrictions on the surface of many of the veins due to the presence of valves. These valves are formed of semilunar folds of the lining membrane and are arranged in pairs. They serve to prevent the blood, whose circulation in the veins is sluggish, from flowing back.

There are two sets of veins, the superficial and the deep, which communicate with each other. In fact, all the veins, large and small, anastomose very freely, especially in the skull and neck, where obstruction would result in serious trouble, throughout the spinal cord, and in the abdomen and pelvis. The deep veins accompany the arteries in their sheath, while the superficial ones have thicker walls and run between the layers of the superficial fascia under the skin, terminating in the deep veins. In the skull the venous channels take the form of sinuses, formed by a separating of the layers of the dura mater, with an endothelial lining that is continuous with that of the veins.

The _capillaries_ are intermediate between the arteries and the veins, the final division of the arteries and the first source of the veins. They are tiny vessels with but a single coat, continuous with the innermost coat of both arteries and veins and consisting practically of one layer of cells with a small amount of connective tissue between. They spread in a great network throughout the tissues, forming plexuses and being especially abundant where the blood is needed for other purposes than local nutrition, as in the secreting glands. Their diameter is so small that the red corpuscles have to pass in single file and may even then be squeezed out of shape. As they have no muscular tissue in their walls, they have no power of contracting. Their walls, however, like those of the smaller arteries and veins, are porous and by virtue of this quality they play an important part in the economy, since in them the exchange takes place between the tissues and the blood.

The arteries in general carry freshly oxidized blood and the veins blood from which the oxygen has been largely used up and which contains waste material. In the pulmonary system, however, the case is reversed, the pulmonary arteries conveying venous blood, as it is called, from the heart to the lungs to be oxidized and the veins returning the blood after it has received its new supply of oxygen.

The pumping of the blood through the arteries is assisted by the contractions of the muscular coat, while the elastic tissue, of which it contains a certain amount, gives elasticity to the walls and enables them to stretch and so to accommodate the larger blood supply forced into them at each beat by the heart. The walls of the veins have not the power of contracting and the blood is pushed through more by gravity and the action of the arteries than by any action of their own.

The walls of all the vessels are nourished by tiny blood-vessels in the outer coat, known as *vasa vasorum*, and the nerves that regulate the action of the arteries are the vasomotor nerves from the vasomotor center in the medulla. Sufficient impulse goes from this center to the blood-vessels all the time to keep them somewhat contracted, in a state of tone, that is, which is increased or diminished as the blood supply is to be diminished or increased.

Lymphatic System.—The lymphatic system also extends throughout the body and consists of a system of channels, spaces, and glands very closely related to the circulatory system and containing a fluid called lymph. There are three principal parts to the system: 1. the lymph spaces, which are open spaces, with no definite walls, in the connective tissue framework of the body, more frequent near arteries and veins and especially so among the capillaries; 2. the lymph capillaries or small vessels which connect the lymph spaces; and 3. the lymphatic vessels, of which there is a deep and a superficial set, the latter accompanying the superficial veins on the surface of the body, the former accompanying the deep blood-vessels.

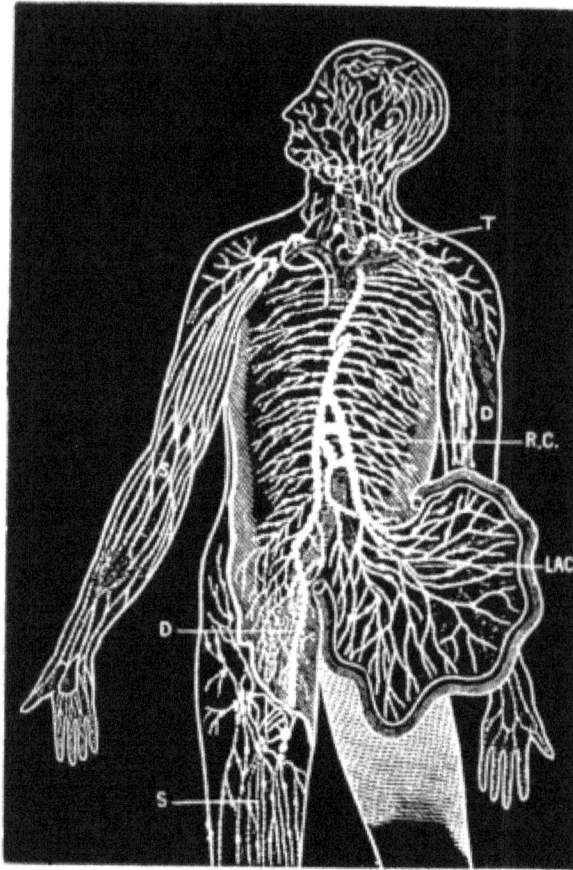

FIG. 7.—Diagram showing the course of the main trunks of the absorbent system: the lymphatics of lower extremities (D) meet the lacteals of the intestines (LAC) at the receptaculum chyli (R.C.), where the thoracic duct begins. The superficial vessels are shown in the diagram on the right arm and leg (S), and the deeper ones on the left arm (D). The glands are here and there shown in groups. The small right duct opens into the veins on the right side. The thoracic duct opens into the union of the great veins of the left side of the neck (T). (Yeo.)

The *lymph spaces* are generally small, though there are some large serous cavities, such as the abdomen, that may be considered as extended lymph spaces.

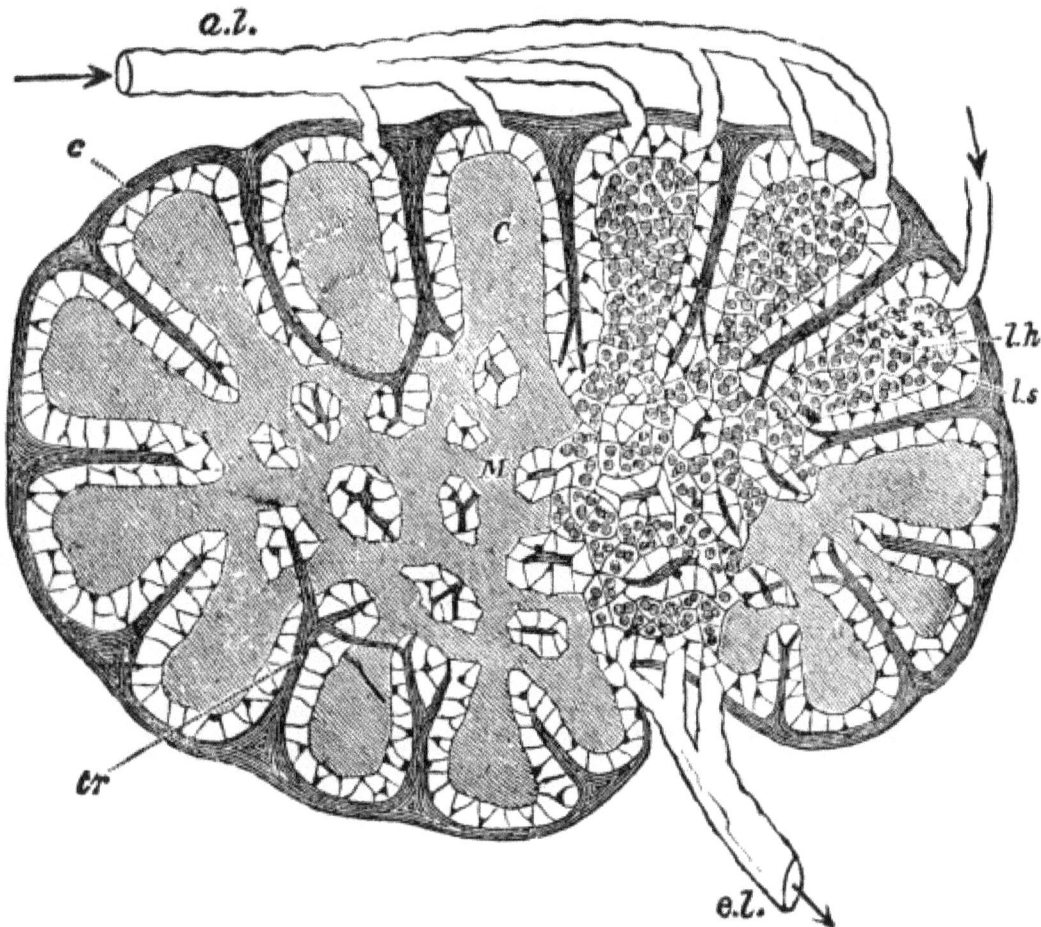

FIG. 8.—Diagram of a lymphatic gland, showing afferent (*a. l.*) and efferent (*e. l.*) lymphatic vessels; cortical substance (*C*); medullary substance (*M*); fibrous coat (*c*); sending trabeculæ (*tr*) into the substance of the gland, where they branch, and in the medullary part form a reticulum; the trabeculæ are surrounded by the lymph path or sinus (*l. s.*), which separates them from the adenoid tissue (*l. h.*). (Sharpey.)

The *lymphatic vessels* have delicate, transparent walls, with three coats like the arteries, though much thinner, and anastomose even more freely than the veins. They have a beaded appearance due to the presence of numerous valves, which form constrictions on their surface. The right lymphatic duct, which is only about an inch long, drains all the lymphatics of the right half of the upper part of the trunk, the head, and the neck approximately, while the thoracic duct drains those of the rest of the body. The latter, which is the largest vessel of the system, begins opposite the second lumbar vertebra with a bulb-like reservoir for the lymph or chyle, the receptaculum chyli, and extends up along the spinal column for a distance of about eighteen inches to the seventh cervical vertebra, where, with the right lymphatic duct, it empties into the left subclavian vein at its junction with the internal jugular, thus establishing direct communication between the lymph spaces and the venous system. The orifices of both vessels are guarded by semilunar valves to prevent regurgitation of the blood.

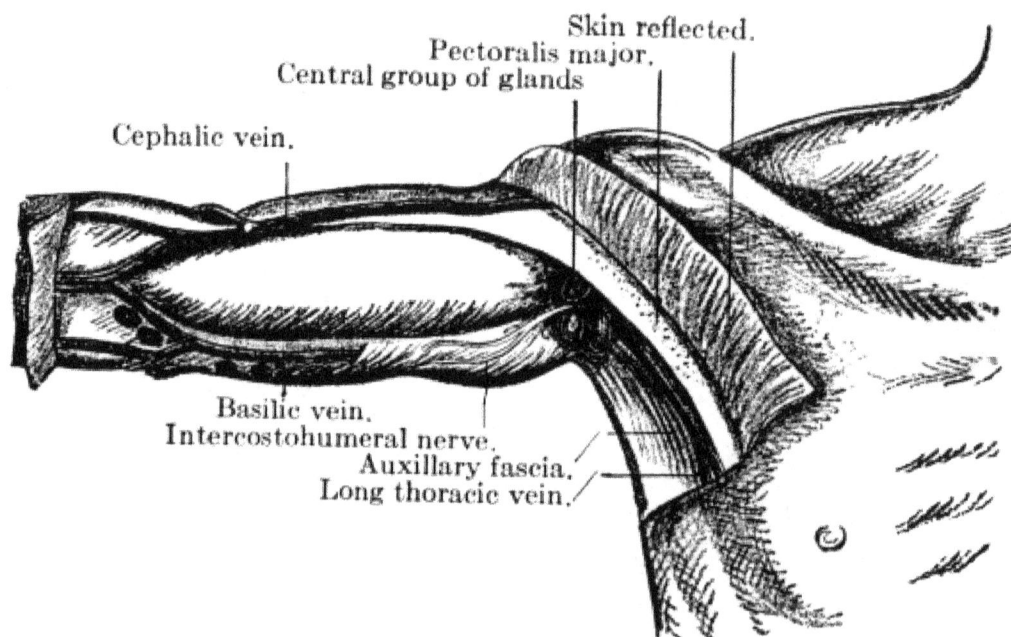

FIG. 9.—Central (superficial) lymphatic glands of the axilla. (After Leaf.)

The _lymphatic glands_ are small oval glandular bodies and occur here and there along the course of the lymphatics. Before entering one of them the vessel breaks up into several afferent vessels which form a plexus within and then emerge again as several efferent vessels which soon unite to form one trunk. These glands occur chiefly in the mesentery, along the great vessels, and in the mediastinum, axilla, neck, elbow, groin, and popliteal space.

The _lymph_ varies in character with the locality, being a little thicker and more opalescent in the lacteals, as the lymphatics of the small intestine are called, especially during digestion, when fat is present. Here it is called chyle. Otherwise it is generally a clear, transparent and slightly opalescent fluid, which, owing to the presence of fibrin, clots when drawn from the body and allowed to stand. In fact, it resembles blood plasma very closely in composition and, as it also contains a certain number of corpuscles or leucocytes that just correspond to the white corpuscles of the blood, it is practically blood without the red corpuscles. These leucocytes have considerable power of amœboid movement and are thought by some to play an important part in the absorption of food.

Owing to intracapillary pressure, the lymph transudes into the lymph spaces and bathes the tissues, being carried away again by the lymphatics. The amount of transudation is determined by the blood pressure—the greater the pressure, the greater the amount of transudation—and is increased by some organic action of the cells in the walls of the vessels. In the process of transudation a certain amount of solid matter goes through the wall of the vessel and it is probable that certain protein elements can be carried thus from the blood-vessels to the lymphatics, though they do not pass through the capillary wall as readily as other substances. Some lymph is also probably formed by the action of the tissues themselves, though the process is not understood.

All muscular movements, active or passive, including the respiratory movements, tend to drive the lymph on its way by pressure, the valves of the vessels keeping it from flowing back. Moreover, its flow is from the capillaries to the veins or from a region of high pressure to one of less pressure. There is probably also some contraction in the walls of the vessels themselves, and the continual formation of lymph helps to drive it along. If an obstruction to the circulation occurs, however, back-pressure results and causes too great transudation. In that event a limb becomes swollen, pale, and generally cool. It pits on pressure, the pressure driving the lymph out and there being no circulation to bring it back. This condition is called œdema and occurs in liver, kidney, and heart troubles, being generally first observed at the ankles. In ascites, hydrothorax, hydrocephalus, and pericardial and pleural effusions the fluid corresponds to lymph in its composition and the large amount is due to excessive formation of the fluid, which is normally present in small quantities.

Lymph gives the tissues substances from the blood that they need and carries off those they do not, whether waste or substances of use to other tissues. Because they thus absorb certain materials not needed by the tissues and convey them to the circulation, the lymphatics have also been called absorbents. Indeed, lymph may be spoken of as the middleman between the blood and the tissues.

Another function of the lymph is to lubricate. Thus, the synovial fluid of the joints is lymph and the pleuræ and the pericardium contain lymph or serum to reduce the friction between the adjoining surfaces as much as possible. The brain and spinal cord do not quite fill the cavities of the cranium and the spinal column but float on a cushion of lymph, the cerebro-spinal fluid. When the brain, which is subject to increase and diminution in size, increases in size, it drives the lymph out, and when it diminishes, the lymph returns.

The lymph glands serve as a protection to adjacent parts and when it leaves the gland the lymph is purer and richer in leucocytes than when it entered. In fact, they filter harmful matter from the lymph and apparently also form white corpuscles. Normally they can with difficulty be felt, but in disease, if the leucocytes are unable to destroy or carry off the poison, the lymph carries it along to the glands, which swell and become tender. If the infection is not severe the swelling goes down and the tenderness passes after a short time, but if it is severe, there may be suppuration and abscess formation and the gland even perhaps be destroyed, giving its life for the health of the part. Thus a wound in the foot, if infected, may cause irritation and enlargement of the glands at the knee and in the groin.

The lymphatic glands are frequently the seat of tubercular infection, especially in the neck, and are enlarged in scarlet fever, tonsillitis, and diphtheria. In syphilis there is general glandular enlargement, and the glands in the groin become enlarged in all diseases of the genital organs. In malignant growths, such as cancer, the extension of the disease is often along the lines of the lymphatics.

Glands.—Of glands in general a word might now be spoken. They are of two kinds, excreting and secreting, and, when simple, are formed by the folding in of a free surface, as in the case of the salivary, gastric, and sebaceous glands, the cells at the gland becoming so modified as to be able to perform the function of excreting or secreting. In racemose glands the gland is broken up into many pockets. Excreting glands take from an organ or from a part substances which have outlived their usefulness and are to be cast out of the body, while the secreting glands form from the blood substances that did not exist in it before, but which are of use to the body, as the ptyalin of the saliva. A strict line cannot, however, be drawn between the two kinds of glands, most glands partaking more or less of both functions, though the sebaceous and sweat glands are probably purely excreting glands and the salivary glands are almost purely secreting. The glands, moreover, are more or less interchangeable in their functions, that is, they have vicarious function, and one gland can take up and do for another what that other is for some reason unable to do. In jaundice, where there is stoppage of the bile duct, the kidneys help out the liver by excreting the bile. If one kidney is removed the other does work for both, and the glands of the skin may help out the kidneys or *vice versa*. Hemorrhage from the lungs sometimes occurs in suppression of the menses.

In a general way the *function of glands* is chemical. They filter out by osmosis, selecting the useful parts for secretion and the useless for excretion. In the chemical action that goes on considerable energy is given off, as is shown by the amount of pressure in the glands and by the fact that their temperature is higher than that of the blood. They all work in a reflex manner, being under the control of the central nervous system. Thus, what is eaten affects the nerve terminals in the mouth, the sensation passes to the nervous system, and an impulse is carried by the motor nerves to the salivary glands.

Most of the glands have ducts to convey away their secretion to other parts of the body or to send excretions out of the body, but there are also *ductless glands*, which, though they seem to have some important function in the process of metabolism, are not well understood. Most of them seem to manufacture some substance that is absorbed by the tissues and that plays an important part in the bodily metabolism, though nothing is secreted by them externally. They are said to have an internal secretion, whereas the glands with ducts have an external secretion. The liver has both forms of secretion, the bile which is sent out and the glycogen that is stored. The ductless glands are the thymus and thyroid glands, the suprarenal capsules, and the pituitary body in the brain.

Nervous Tissue.—Presiding over all the organs, muscles, and blood-vessels, as the source of all action and all sensation, are the nerves. Nervous tissue is of two kinds: 1. the gray or vesicular, which originates impulses and receives impressions, and 2. the white or fibrous, which conveys impressions. The gray matter consists of large granular cells of protoplasm containing nuclei, which give off many branches or dendrites. From the under surface there usually comes one main branch, the axis-cylinder process. These processes sometimes give off branches and sometimes not, but they form the nerve fibers and carry impulses away from the nerve cells. The cells of the processes are elongated in shape, have a nucleus, and are placed end to end, with a definite constriction between them.

Each axis-cylinder process is surrounded by a sheath called the medullary sheath, while each nerve fiber consists of a central axis-cylinder process surrounded by the white substance of Schwann and enclosed in a sheath. A bundle of these fibers invested in a fibro-areolar membrane called the neurilemma constitutes a nerve, and of these the white matter is formed. The blood supply is brought by minute vessels, the *vasa nervorum*.

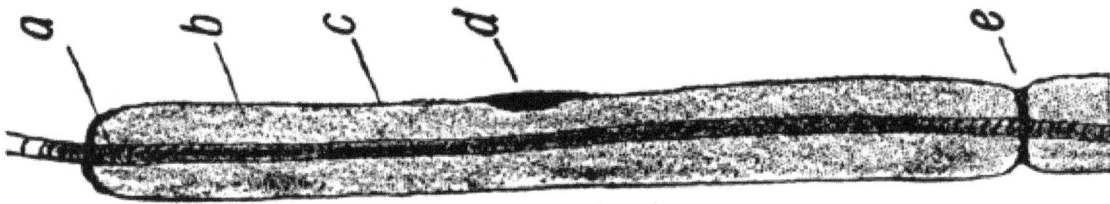

FIG. 10.—Longitudinal nerve fiber (diagrammatic): *a*, Axis-cylinder; *b*, medullary sheath; *c*, neurilemma; *d*, nucleus; *e*, node of Ranvier. (Leroy.)

The nerves of the cerebro-spinal system preside over animal life and have to do with voluntary acts, while those from the sympathetic system regulate organic life and are quite independent of the will. Both sensory and motor nerves extend all over the body, accompanying the arteries in a general way. The sensory nerves end on the surface in plexuses, in end bulbs situated in the papillæ of the skin, or in tactile corpuscles, these last occurring more especially where there is no hair. The motor nerves end peripherally in plexuses or by end plates. The central terminations of the motor nerves and the terminations of sensory nerves in special organs, except where they end in a cell, are not well understood.

Like muscles, nerves are probably never at rest, for through them the muscles get their tone. When a nerve acts, no heat is produced and there is no change in the nerve afterward, as there is in muscle. Probably nerve impulse is the transmission of physical rather than chemical changes along the fiber, the atoms of the nerve being set in vibration and the vibrations being transmitted along its length. Stimulation is produced by physical injury, by chemical influence, by electricity, by heat, and the message is always referred to the nerve termination. Thus, if the nerve at the elbow, over the "crazy bone," is touched, a tingling is felt in the fingers rather than at the point of pressure. A person who has had an arm or leg amputated will frequently speak of his fingers or toes on that side being cold, or complain of pain in them, because the scar below the point of amputation tightens around the nerves and pinches them.

It is through the nerves that people get in touch with the outer world and that they judge of size, weight, etc. All careful adjustment of the muscles is under the control of the nervous system.

CHAPTER II.
THE SKIN, ITS APPENDAGES AND ITS FUNCTION.

The whole exterior surface of the body is covered by the skin, an excreting and absorbing organ, which serves as a protection to the parts beneath and is also the organ of touch. It has two layers, a superficial and a deep. The superficial layer, the epidermis or cuticle, is composed wholly of epithelial cells, of which the deepest layer is columnar and moulded upon the papillary layer of the derma, while the intermediate layers are more rounded and the surface ones flat. The deepest layer also contains the skin pigment, which causes the variation in shade between the Indian, the negro, and the white man. Below the epidermis, which is chiefly protective, is the tough, elastic, and flexible tissue of the derma or true skin, in which are vested most of the activities of the skin. Its surface is covered with papillæ, which are more numerous in the more sensitive parts. Each papilla contains one or more capillary loops and one or more nerve fibers, while some terminate in an oval body known as a tactile corpuscle. Beneath the papillæ is the reticular layer, composed of interlacing bands of fibrous tissue and containing blood-vessels, lymphatics, and nerves, as well as unstriped muscle fibers where hair is present.

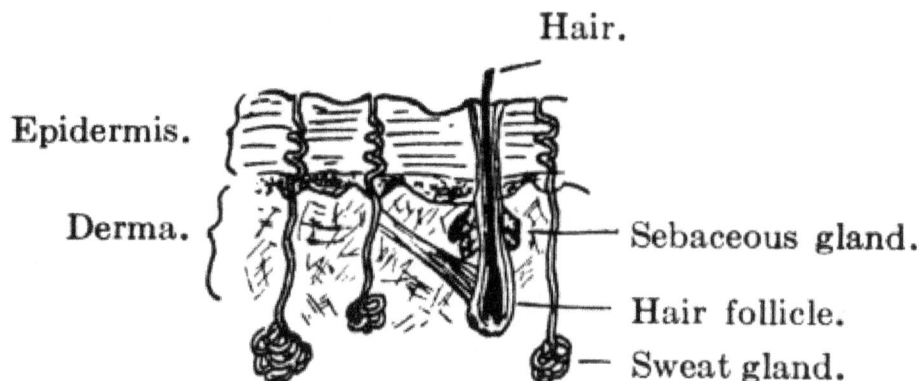

FIG. 11.—Vertical section of skin.

At the apertures of the body the skin stops and is replaced by *mucous membrane,* an integument of greater delicacy but which consists fundamentally of the same two layers, a superficial, bloodless epithelium and a deep fibrous derma. It is continuous with the skin, but is much redder and more sensitive and bleeds more easily. The passages and cavities that it lines, unlike those lined by serous membranes, communicate with the exterior of the body and are for that reason protected against contact with foreign substances by mucus, which is thicker and more sticky than the lymph that moistens the endothelium found on serous surfaces. Mucous membrane is found in the alimentary canal, the respiratory tract, and the genito-urinary tract. In cavities, like the stomach and intestines, which are subject to variations in capacity, it is thrown into folds or rugæ. The mucus is secreted by small glands in the membrane.

Appendages of the Skin.—The skin has various appendages. On the dorsal surface of the last phalanges of the fingers and toes are flattened and horny modifications of epithelium, the *nails.* They have a root embedded in a groove of skin by which they grow in length and a vascular matrix of derma beneath them which gives growth in thickness. To their growth in length there seems to be no limit.

The *hairs* also, which occur all over the body, except on the palms of the hands and the soles of the feet, are a modification of the epithelium. Each hair has a bulbous root springing from an involution in the epidermis and derma called the *hair follicle,* into which one or two sebaceous glands empty. It is raised by involuntary muscle fibers and grows by constant additions to the surface by which it is attached. This growth seems, however, to be limited, and when its term is reached the hair falls out and is replaced by another. The horny epithelial cells that go to form the hair contain the pigment that gives it its color.

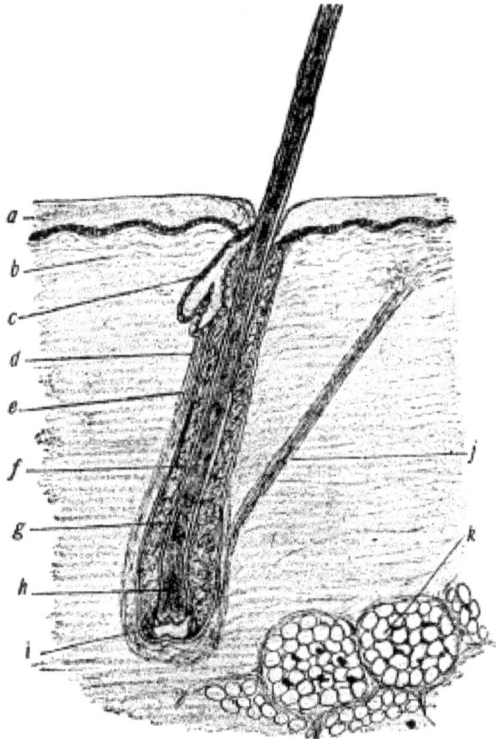

Fig. 12.—Skin and longitudinal section of hair: *a*, Epidermis; *b*, corium; *c*, sebaceous gland; *d*, fibrous root-sheath; *e*, glassy membrane; *f*, outer root-sheath; *g*, inner root-sheath; *h*, expanded bulbous end of hair; *i*, papilla of hair; *j*, arrector pili; *k*, adipose tissue. (Leroy)

Like the hairs, the *sebaceous glands* are situated in all parts of the body except the palms of the hands and the soles of the feet. They lie in the papillary layer and empty into the hair follicles, except occasionally, when they empty directly upon the surface of the skin. They secrete an oily substance, *sebum*, the débris resulting from the degeneration of the epithelial cells of the gland itself, which serves to keep the hair glossy and the skin soft and flexible.

The *sweat glands*, on the other hand, are more frequent on the palms and soles and though sometimes found in the derma are usually situated lower down in the subcutaneous cellular tissue. They are least numerous on the back and neck. Coiled up in the lower layers of the skin, they discharge the sweat through a spiral excretory duct upon its free surface.

The *sweat* is a clear, colorless, watery fluid with a salty taste, an alkaline reaction, and a characteristic odor that varies with the individual. If very scanty, it may be acid in reaction. Besides water it contains a small percentage of solids, as inorganic salts, especially sodium chloride, fatty acids, neutral fats, and at times, especially in some diseases of the kidneys, urea, that is, the end-products of the metabolism of starches and fats chiefly. There is usually also some carbon dioxide, whence the expression cutaneous respiration.

The sweat serves to keep the skin moist and in good condition, to remove outworn and poisonous or irritating matters, and to regulate the temperature. As a rule it evaporates upon reaching the surface, in which case it is known as invisible or insensible perspiration, but if conditions of the atmosphere are not favorable to prompt evaporation, as when the air is damp, the skin becomes damp and there is visible perspiration.

Though an abundant supply of blood increases the action of the sweat glands, they are regulated by definite secretory nerves rather than by the vasomotor nerves. In a cold sweat the action is probably due to some disturbance of the nerve supply without increase of the blood supply. Ordinarily perspiring is a reflex act due to the stimulation of the afferent cutaneous nerves, as by the application of heat, but sometimes, as in cases of strong emotions, involuntary impulses are sent from the brain to the spinal centers and so arouse the action of the glands. Atropin has the power of preventing the secretion of sweat by paralyzing the terminations of the secretory nerves, while pilocarpin produces an opposite effect in a similar way.

On account of these sweat glands the skin becomes next in importance after the kidneys in the excretion of waste products. The quantity of sweat excreted varies greatly and is hard to measure. It is influenced by the temperature and humidity of the surrounding air, by the nature and quantity of food and drink consumed, by the amount of exercise, the relative activity of other organs, especially the kidneys, and by certain mental conditions. The hotter it is, the greater the amount of perspiration. In damp weather there may be less perspiration, but it does not evaporate and is therefore more in evidence.

Ordinarily man has a temperature of 98.6°. The source of this body heat or temperature is the general body metabolism, muscular activity, and activity of the glands, especially of the liver, which is constantly active, the blood in the hepatic vein being warmer than that in any other part of the body. The tissue of the brain also is said to be warmer than the surrounding blood, and the heart and respiratory muscles, which are in constant activity, are responsible for much of the body heat. The amount of heat generated in the body, therefore, varies at different times, according as a person is awake or asleep, quiet or active.

Temperature Regulation.—The temperature is regulated by variations in the production and loss of heat, less being known of its production than of its loss. It has been calculated that four-fifths of the energy of the body is converted into heat, one-fifth into work. As the minimum amount of heat produced in twenty-four hours is sufficient to raise 10 gallons of water from 0° to boiling-point, it is evident that if there were not some way for the escape of much of this heat the body would become hotter and hotter and finally destroy itself. The temperature, however, except on the surface, is uniform, heat being lost as fast as it is produced. For, although oxidation at any point raises the heat of the blood at the point, this heat is carried by the blood to other parts, to which the surplus is given up, while blood cooled in the skin goes to the hotter inward parts to cool them and be warmed itself. In fact, heat is expended by conduction and radiation, through respiration, perspiration, and heat given to the urine and fæces. It is, therefore, largely, 75 to 80 per cent., carried off through the skin and the lungs; 60 to 70 per cent. is lost by radiation to the air and other bodies with which the body comes in contact; 20 to 30 per cent. is lost by the evaporation of sweat, 4 to 8 per cent. by the warming of expired air, urine and feces, and 1 to 2 per cent. by cold food that is taken in. Radiation acts more favorably where the surroundings are cool and the air in motion, as on a breezy day. Conduction is carried on best where the surrounding air is cool, especially if it is moist, for moist air is a better conductor of heat than dry air. Evaporation is very important in hot weather or where men work in hot air.

Even in health the temperature may range from 98.6° to 99.5°, and a degree or two below or above is not dangerous. When a person first gets up in the morning his temperature is apt to be subnormal, but after food and exercise have been taken it becomes normal and stays so till the end of the day, when, if the person is tired, it may go up a little. If a person is tired out, the temperature is apt to be subnormal. There is also in the body what is called the vital tide, which is highest afternoon and evening and lowest in the morning.

The rate of production of heat varies greatly in different people. One person uses a certain amount of tissue more quickly than another, that is, he lives faster. Moreover, size makes a difference in that a small body has more surface to its weight than a large one and so has to produce the same amount of heat at a faster rate in order to maintain the right temperature. Taking food increases heat, probably because of the muscular effort needed to eat it. Muscular work is another factor. And finally the whole matter of heat production seems to be under the control of the nervous system. Not much is known on this point except that there is a heat center in the medulla which plays an important part in heat production and whose influence is seen where the temperature shoots way up in disease just before death. It is now thought that fever is due to a disturbance of this nervous mechanism, though just what the disturbance is is not known.

Fever is a condition of increased bodily temperature, due to increased production or to decreased loss of heat. As a rule, in all fevers the metabolic changes in the body are increased. Hence the patient becomes emaciated in a long fever. The frequent increase in the amount of urea during fever shows an increase in protein metabolism. The temperature in fevers rises as high as 106° and in sunstroke sometimes to 110°. Except in sunstroke a higher temperature than 106° generally means death. Subnormal temperature is due to a decrease in the bodily metabolism and so to lessened heat production. As a rule, if the functions are all active, especially that of the sweat glands, a person can be exposed to severe heat without the temperature being affected, though sometimes on a hot summer day it may be up half to one degree. The cause of heat-stroke with its high fever is unknown, but probably it is due to some effect on the heat center in the brain. Heat prostration is also due to prolonged exposure to heat, but is generally accompanied by a subnormal temperature. The effect of cold, as in freezing, is to diminish all the metabolic activities of the body. The temperature can be artificially regulated more or less by variations of food, varying amounts of exercise, by drugs, etc.

Sense of Touch.—Before passing on to a discussion of the individual parts, a few words might well be said of the sense of touch, since that is general and resides largely in the skin, whose other functions have just been described. It may be regarded as the form from which all the other special senses have developed, certain portions of the body having become more sensitive than others to certain vibrations, as the eye to those of light. The internal organs probably have little sense of touch.

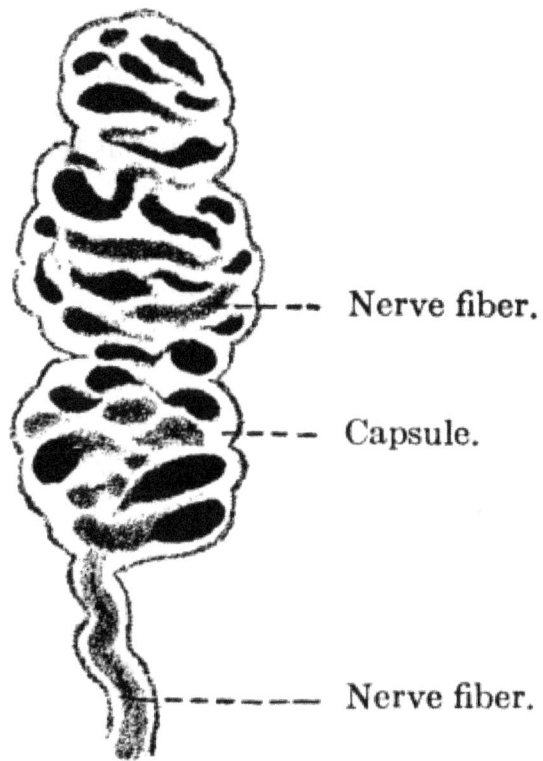

- - Nerve fiber.

- - - Capsule.

- - - - - - Nerve fiber.

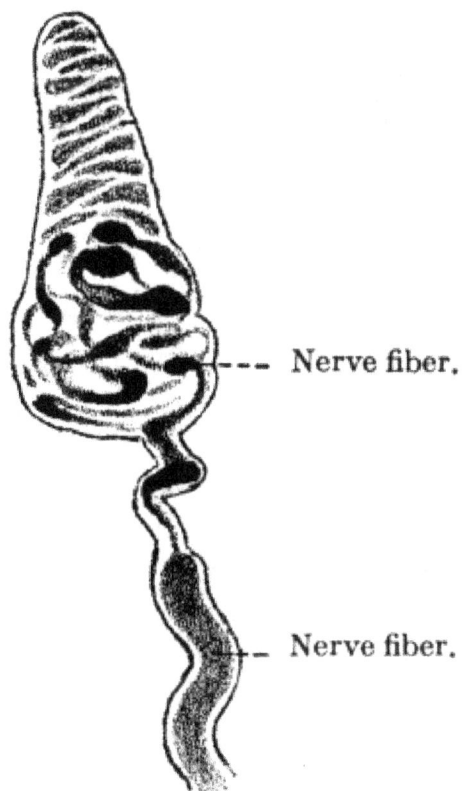

- - - Nerve fiber.

- - Nerve fiber.

FIGS. 13, 14.—Meissner's corpuscle from man; ×750.
(Böhm, Davidoff, and Huber.)

Touch is useful only within arm's reach but there gives one a sense of space that sight does not give. It is practically determined by the *touch corpuscles*, which are found in the skin over almost the entire body, though they are more numerous in some places than in others, the distribution of the corpuscles determining the sensitiveness of the skin. These touch corpuscles are protoplasmic bodies containing nuclei, about which are entwined filaments from the cutaneous nerves. Where the corpuscles are absent the filaments of the cutaneous nerves themselves play an important part. The finger tips have a very delicate sense of touch and the tip of the tongue is the most sensitive part of the body. Hence spaces in the mouth seem larger than elsewhere. By the transmission of sensations of touch to the brain the sensation is localized and the tactile sensation becomes a tactile perception.

There are three main divisions of the sense of touch: 1. sensations of touch proper or tactile sensation; 2. sensations of temperature, and 3. sensations of pain. The temperature sense is the transmission by the skin of sensations not so much of a certain degree of heat or cold as of the difference between the temperature of an object and that of the skin. The longer an object is in contact with the skin, the less conscious the person is of it, not only because it becomes of the same temperature, but also because he becomes accustomed to it. There also seem to be in the skin, besides the touch corpuscles, two other terminal organs with separate nerve fibers, the one for detecting heat, the other cold; for there are places on the body where heat can be detected and cold cannot, and *vice versa*.

Sensations of pain may be merely an exaggeration of tactile sensation, as in too hard pressure or too great heat, but there seems to be also a sensation of pain in the skin. All organs are said to have common sensibility to pain and any exaggeration of this sensibility causes a sensation of pain. All the special senses require a certain amount of judgment in the interpretation of the sensations they convey.

CHAPTER III.
THE CRANIUM AND FACE.

The intelligence and all the special senses, except the sense of touch already spoken of, are gathered together compactly in the head, where they are carefully protected with bony tissue. Covering the brain is the skull or cranium, which is made up of eight bones, the frontal, the occipital, two parietal, two temporal, the sphenoid, and the ethmoid, while the bones of the face are fourteen in number, two nasal, two superior maxillary, two lachrymal, two malar, two palate, two inferior turbinated, the vomer, and the inferior maxillary. For the most part the bones are arranged in pairs, one on either side.

The Cranial Bones.—The cranium or skull is especially adapted for the protection of the brain and the bones are flat and closely fitted to its surface. They have two layers of bone, the outer and the inner tables, of which the outer is the thicker, and between these is a tissue filled with blood-vessels, the *diploë*. In the infant, whose brain has not yet attained its full size, opportunity must be left for growth and the skull therefore consists of a number of bones with interlocking notched edges, where growth takes place, but in the adult it forms one solid covering of bone.

The line where the edges of two cranial bones come together is called a *suture*. The suture between the frontal bone and the forward edges of the two parietal bones is called the *coronal suture*, that between the two parietal bones at the vertex of the skull is known as the *longitudinal* or *sagittal suture*, and that between the occipital bone and the back edges of the parietal bones as the *lambdoidal suture*.

Where the coronal and sagittal sutures meet is a membranous interval known as the *anterior fontanelle*, while the *posterior fontanelle* is at the juncture of the sagittal with the lambdoidal suture. These fontanelles—so called from the pulsations of the brain that can be seen in them—close after birth either by the extension of the surrounding bones or by the development in them of small bones known as *Wormian bones*, the posterior one closing within a few months, the anterior by the end of the second year. In rickets, however, the anterior fontanelle remains open a long time, sometimes into the fourth year.

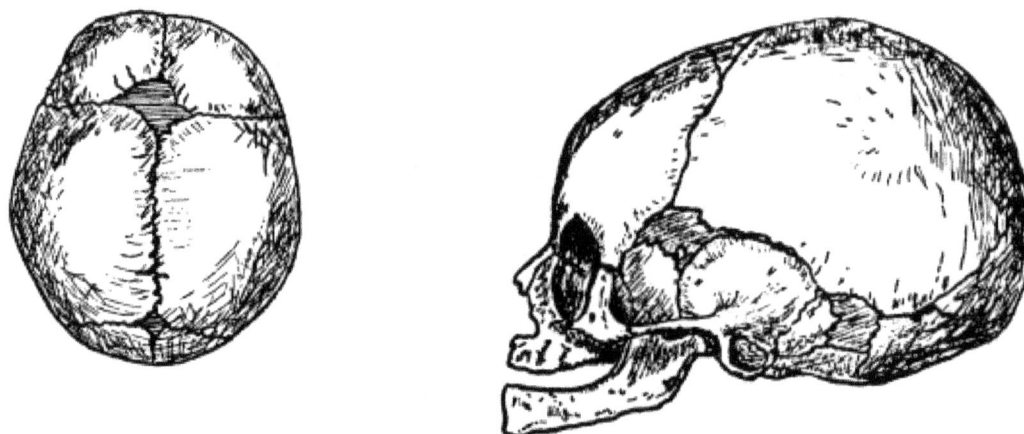

FIG. 15.—Cranium at birth, showing sutures and fontanelles.

The <u>frontal bone</u>, as its name implies, forms the fore part of the head or forehead. It joins the parietal bones above and the temporal bones on either side. At the lower edge are the supra-orbital arches, each with a supra-orbital notch or foramen on its inner margin for the passage of the supra-orbital vessels and nerve, the nerve most affected in neuralgia. Just above the arches on either side are the superciliary ridges, behind which, between the two tables of the skull, lie the frontal sinuses. On the inner surface the frontal sulcus for the longitudinal sinus runs along the median line.

The **parietal bones** are the side bones of the skull. They meet each other in the sagittal suture at the median line above and join the frontal and occipital bones at either end, while below they touch upon the temporal bones, the temporal muscles being attached in part along their lower surface. These muscles are inserted into the coronoid process of the lower jaw, which they thus help to raise and to retract.

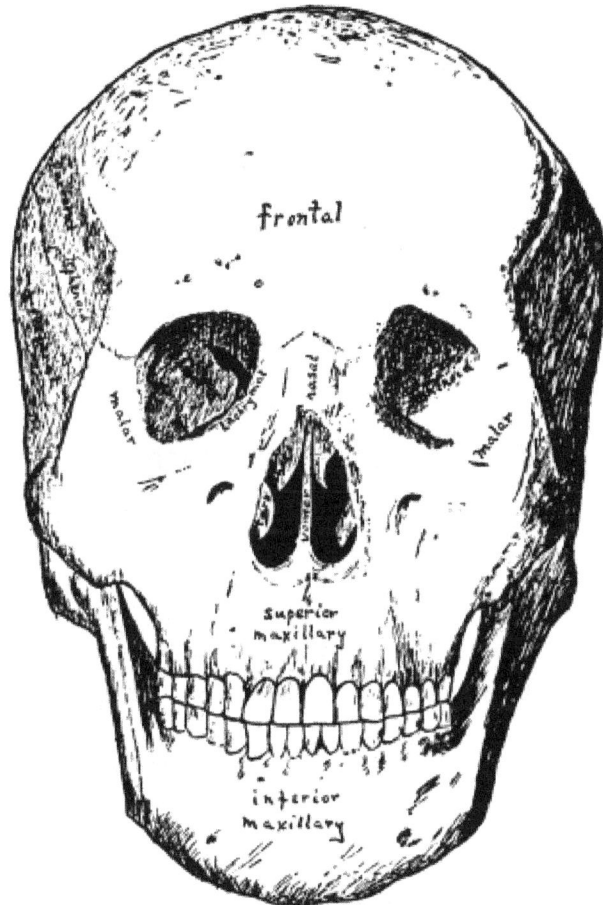

FIG. 16.—Front view of the skull.
(After Sobotta.)

The **occipital bone** is at the base of the skull and at birth consists of four pieces. In the lower, anterior part is the foramen magnum, an oval opening through which the spinal cord passes from the skull down into the spinal canal. Half way between the foramen and the top of the bone is the external occipital protuberance for the attachment of the ligamentum nuchæ which holds the head erect. The inner side of the bone is deeply concave and is divided by a cross-shaped grooved ridge into four fossæ, the internal occipital protuberance being situated where the arms of the cross meet. The occipital lobes of the cerebrum lie in the two upper fossæ and the hemispheres of the cerebellum in the two lower ones. In the grooves upon the ridge are the sinuses which collect the blood from the brain.

The occipital and frontal muscles, united by a thin aponeurosis, cover the whole upper cranium and are known as the *occipito-frontalis muscle*. At the back this is attached to the occipital bone, while in front it interlaces with various face muscles. It is a powerful muscle and raises the brows, wrinkles the forehead, and draws the scalp forward. Long hair grows on the skin over it as a further protection against blows upon the skull and sudden variations in temperature.

The **temporal bones**—said to be so named because the hair over them is the first to turn with age—are situated at the sides and base of the skull and are in three portions: the squamous or scale-like, the mastoid or nipple-like, and the petrous or stony portion. The squamous is the upper portion and has projecting from its lower part the long arched zygomatic process, which articulates with the malar bone of the face and from which arises the *masseter muscle,* one of the chief muscles of mastication, which has its insertion in the ramus and angle of the lower jaw. Just above the zygomatic process the *temporal muscle* has its origin in part, while below is the *glenoid fossa* for articulation with the condyle of the lower jaw, the posterior portion of the fossa being occupied by part of the parotid gland.

The rough mastoid portion of the temporal bone is toward the back and affords attachment to various muscles, of which the most important are the occipito-frontalis and the sterno-cleido-mastoid. Within it are the mastoid cells, which communicate with the inner ear and are lined with mucous membrane continuous with that of the tympanum. They probably have something to do with the hearing. In children they often become the seat of inflammation (mastoid abscess) in infectious diseases and the mastoid bone has to be cut to let out pus that has collected. As the lateral sinus is directly behind the mastoid bone, there is very great danger of going through into the sinus and causing a fatal hemorrhage.

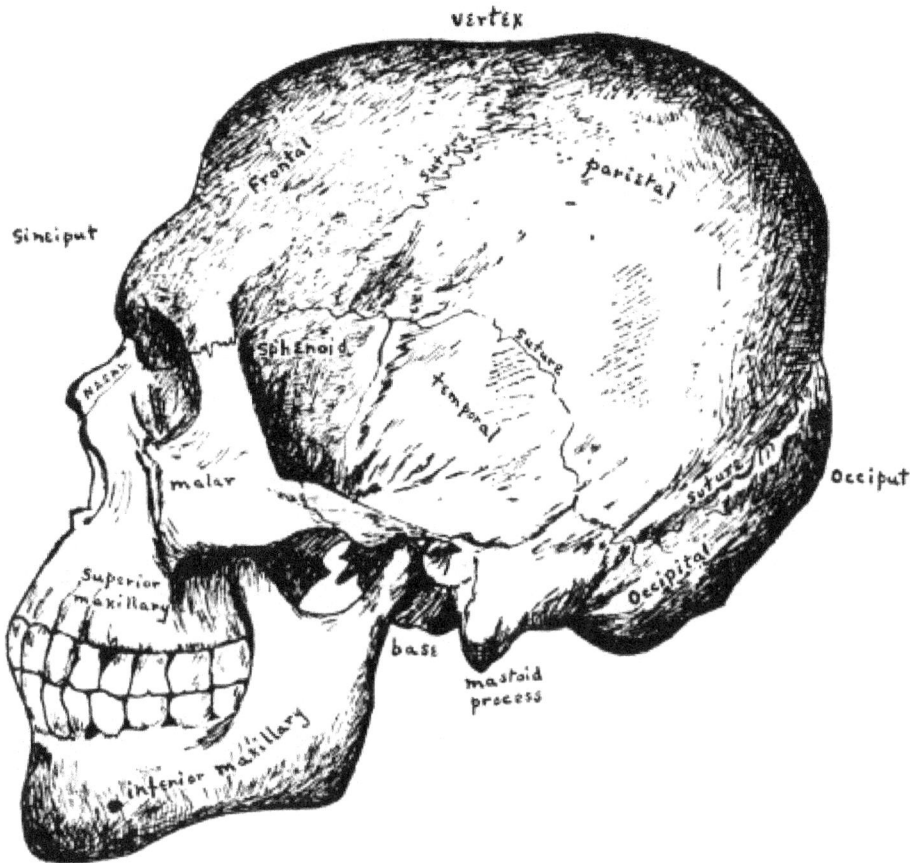

F<small>IG</small>. 17.—Side view of the skull.
(After Sobotta.)

The petrous portion, which contains the organ of hearing, is between and somewhat behind the other two portions, at the lower edge of the temporal bone, wedged between the sphenoid and the occipital bones. On its outer surface is the *external auditory meatus*, and from below projects a long sharp spine called the *styloid process*, to which several minor muscles are attached. In the same angle between the petrous and squamous portions lies the bony Eustachian tube.

The **sphenoid** or **wedge bone**, so called because in the process of development it serves as a wedge, lies at the base of the cranium, forming as it were the anterior part of the floor of the cavity containing the brain. It is a large, bat-shaped bone and articulates with all the cranial and many of the facial bones, binding them all together. It has a body, two large wings, and two lesser wings and, appears on the outside of the skull between the frontal and the temporal bones behind the zygomatic process. In the adult the body of the sphenoid is hollowed out into the sphenoid sinuses, in which pus sometimes forms.

The Ethmoid Bone.—In front of and below the sphenoid and extending forward to the frontal bone is the ethmoid, the last of the cranial bones. It consists of a horizontal cribriform or sieve-like plate, from either side of which depend lateral masses of ethmoid cells. To the inner side of these masses are attached the thin curved *turbinated bones*, superior and middle, while between them is a vertical plate that forms the bony septum of the nose. Rising from the upper surface of the cribriform plate is another vertical plate, the *crista galli*, with the olfactory grooves on either side for the reception of the olfactory bulbs, filaments of the olfactory nerve passing down through the perforations of the cribriform plate to the nose. For the brain, which fills almost the entire cavity of the cranium, is supported by the sphenoid and ethmoid bones internally, as it is protected externally by the other cranial bones.

Ossification of Sutures.—If premature ossification of all the sutures occurs, *idiocy* results, while in *cephalocele* there is a gap in the ossifying of the bones so that the membranes or brain protrude. In *rickets* the forehead is high and square and the face bones poorly developed, so that the head looks larger than it really is. In *Paget's disease* the bones enlarge and soften. This affects the head but not the face and often the first thing noticed is that the hat is too small. *Craniotabes* is thinning of the bone in places, the bone becoming like parchment and being easily bent. It is generally caused by pressure of the pillow or the nurse's arm.

Bones of the Face.—The facial bones serve to form the various features of the face, which after all are merely organs of special sense. Many delicate muscles control the facial expression which, consciously or unconsciously, reflects the character of their owner.

Surgically the most important of the facial bones are the two *superior maxillary bones*, because of the number of diseases to which they are liable. They meet in front, together forming the upper jaw, and with the malar bone help form the lower part of the orbit of the eye. They are cuboid in shape and are hollowed out into a pyramidal cavity called the *antrum of Highmore*, which opens by a small orifice into the middle nasal meatus and which sometimes becomes infected and has to be tapped. The nasal process for articulation with the frontal and nasal bones has, at its lower edge, a crest for the inferior turbinated bone, and close beside this on the inside, extending down from the upper edge, is a deep groove which, with the lachrymal and inferior turbinated bones, helps to form the lachrymal canal for the nasal tear duct. The bones give attachment to many small muscles, connected for the most part with the nose and mouth, of which the masseter is the only important one.

The two *malar* or *cheek bones* are small quadrangular bones, which form the prominences of the cheeks and help form the orbits of the eyes. Projecting backward from each is a zygomatic process for articulation with the zygomatic process of the temporal bone, while a maxillary process extends downward for articulation with the superior maxillary. Here again the most important muscle attached is the masseter. If the malar bone is crushed great deformity results.

The *lachrymal bones* are two small bones, about the size and shape of a finger-nail, situated at the front of the inner wall of the orbit. At the external edge is a groove which lodges the lachrymal sac above and forms part of the lachrymal canal below.

The two *palate bones* are at the back of the nasal fossæ and help to form the floor of the nose, the roof of the mouth, and the orbit. Each has a vertical and a horizontal plate, and it is these last that by their juncture form the *hard palate*. Oftentimes in cases of hare-lip *cleft palate* also occurs, the result of incomplete development. To remedy the consequent opening in the roof of the mouth, which makes articulation difficult, operation is generally resorted to, though sometimes a plate is fitted over the opening by a dentist.

The *nasal bones* are two small oblong bones which articulate with the frontal and superior maxillary bones and with each other. They form the bridge of the nose, the rest of the nose being wholly of cartilage, except for the *vomer*, a bone shaped like a plough-share, which forms part of the nasal septum, articulating along its anterior edge with the ethmoid and the triangular cartilage.

The two *inferior turbinated bones* lie along the outer walls of the nasal fossæ. They are thin scroll-like bones covered with mucous membrane and serve to heat the air as it passes in. Sometimes when one has a cold, the membrane and the bone too swell up and close the nares. Loss of the sense of smell in a bad cold may be due to such swelling and the consequent impeding of the entrance of odoriferous particles—a condition that would likewise interfere with the sense of taste. Part of the bone is sometimes removed, to enlarge the passage, enough being left to warm the air.

Lastly, there is the _inferior maxillary bone_ or _lower jaw_. This has a horseshoe-shaped body and two _rami_, one at either end. Each ramus has a pointed process in front called the _coronoid process_, into which is inserted the temporal muscle. At the back, and separated from the coronoid process by the sigmoid notch, is the _condyle_, which articulates with the glenoid fossa on the temporal bone. The rami also give attachment to the masseter muscle at its point of insertion. In adult age the ramus is almost vertical but in old age the portion of the jaw hollowed out into alveoli for the teeth becomes absorbed and the angle of the jaw becomes very obtuse. On the inner side of the jaw near the middle on either side is the fossa for the _sublingual gland,_ while the _submaxillary gland_ lies in a fossa farther back on either side.

Sometimes the lower jaw is dislocated and when once this has occurred it is liable to occur again, the ligaments becoming stretched.

CHAPTER IV.
THE ORGANS OF SPECIAL SENSE.

The Nose.—The nose, the organ of the sense of smell, is composed of a framework of bones and cartilages, the bridge being formed by the two nasal bones, and the septum by the vomer and the triangular cartilage. It consists of two parts, the *external nose* and the internal or *nasal fossæ*, which open to the face by the anterior nares or nostrils and into the pharynx by the posterior nares. Externally it is covered with skin, internally with ciliated mucous membrane. The *fossæ* have the inferior turbinated bones along their outer walls and are divided into three parts known as the *superior, the middle, and the inferior meatus*, the middle one connecting with the antrum of Highmore, while into the inferior meatus the lachrymal canal empties. There are many small muscles of which little use is made, although in forced respiration, as in pneumonia, where every aid to breathing is called into play, even the alæ nasi or nostrils are made to exert what muscular power they possess in order to supply more air.

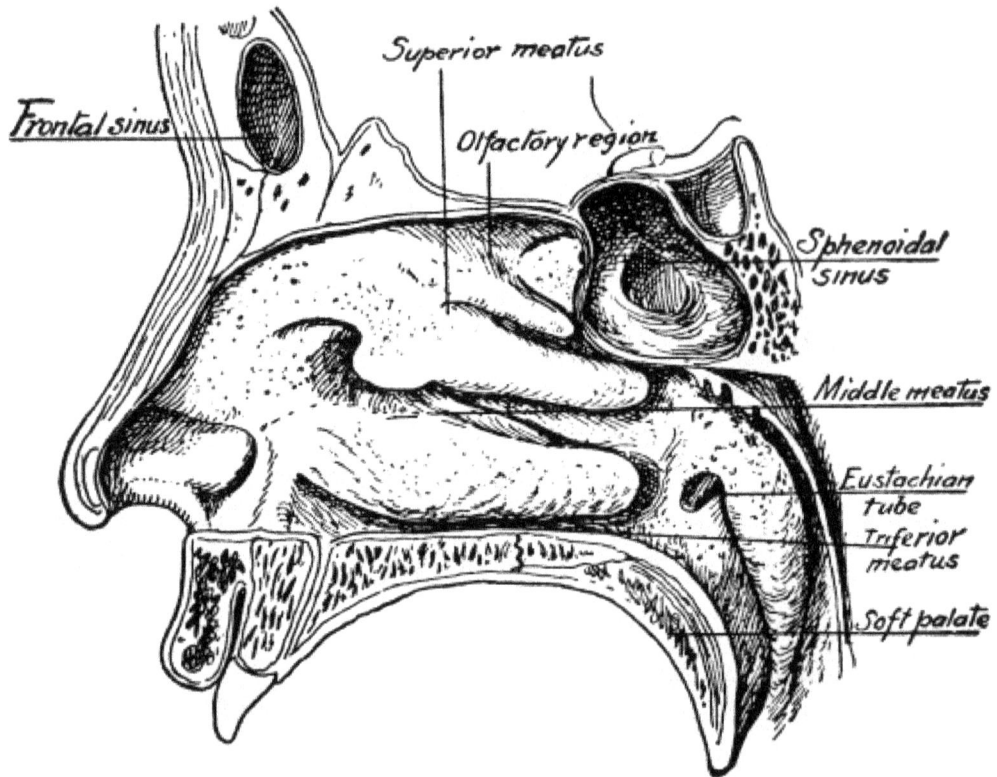

FIG. 18.—The nasal cavity.
(After Sobotta.)

Not only is most of the air breathed in through the nose and warmed in its passage through, but the nose is the organ of smell and by means of the peculiar property of its nerves protects the lungs against deleterious gases and helps the taste discriminate. The *olfactory* or *first cranial nerves*, after emerging from the brain, lie on the under surface of the frontal lobe and rest on the ethmoid bone in what is known as the *olfactory tract*. Each nerve ends in a bulb-like termination called an *olfactory bulb,* which rests on the cribriform plate and sends little terminal fibers down through to be distributed to the nasal cavities, especially to the upper half of the septum of the nose, the roof of the nose, and the anterior and middle turbinated bones. For in the mucous membrane of the upper nasal cavity are specially modified epithelial cells called *olfactory cells*, which play an important part in the conduction of smell. Hence when one wishes to smell anything especially well he sniffs it up.

Probably the *sensation of smell* is caused by odoriferous particles in the atmosphere being breathed into the nose, where they affect the olfactory cells, which transmit the impulses to the olfactory nerve and so to the brain. Whereas a certain amount of moisture in the nasal cavity seems to be essential for accuracy of smell, the presence of too much or too little interferes with it. The mucous membrane has a certain power also of distinguishing different smells at the same time, though this power varies greatly in different people, one smell often wholly overpowering all others.

The cartilage below the bridge of the nose is sometimes attacked in *syphilis* and *cancer*, and *lupus* often begins on the nose. *Deviation of the septum* may occlude all air from one side of the nose, an effect also produced by *polypi*, generally of the turbinated bone. Either condition is easily remedied. *Nosebleed*, though generally unimportant, may be serious in adults.

The Mouth.—The mouth is of great importance as an entrance for fresh air to the lungs when the nasal passages are for any reason impeded and as the resonant chamber from which proceeds the voice, man's chief means of communication with his fellows. Its chief value may be said, however, to reside in the fact that it is the vestibule of the alimentary canal. It is an ovoid cavity lined with mucous membrane and is bounded in front by the *lips*, at the sides by the *cheeks*, below by the floor and tongue, and above by the *hard palate* anteriorly and by the *soft palate* posteriorly, the *uvula* depending from the latter like a curtain between the mouth and the pharynx. Shape is given to the mouth by the bones of the upper and lower jaw and its size is altered by the lowering and raising of the latter, which is quite freely movable.

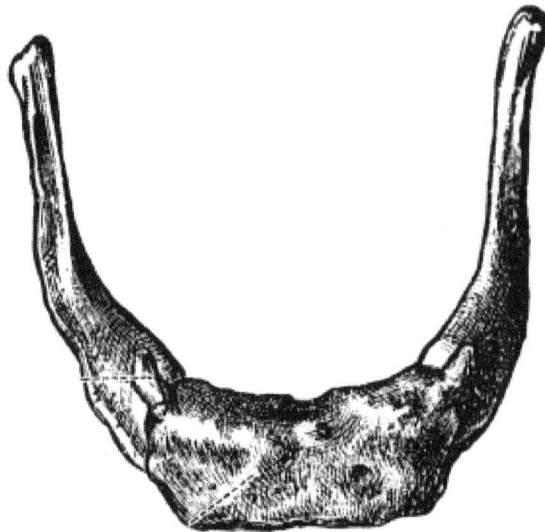

Fig. 19.—The hyoid bone. (Toldt.)

At the back of the mouth, at the entrance to the pharynx, are the *anterior* and *posterior pillars* of the fauces, which contain muscular tissue, and between which on either side are thick masses of lymphoid tissue, the *tonsils*. The floor of the mouth is formed largely by the *tongue*, which completely fills the space within the lower teeth. Its base or root is directed backward and downward and is attached by muscles to the hyoid bone and the lower jaw, the *hyoid bone* being a horseshoe-shaped bone lying just below and as it were within the inferior maxillary. The base of the tongue is attached also to the epiglottis and at the sides to the soft palate by the anterior pillars. Except at its base and the posterior part of its under surface the tongue is free, but a fold of mucous membrane, the *frenum*, holds it somewhat in front. Thus it possesses great versatility of motion and serves as an auxiliary in articulation, mastication, and deglutition.

The Teeth.—Securely embedded in either jaw are the teeth, nature's instrument for the first preparation of the food for digestion through tearing and grinding. The *incisors*, which are in front, have wide sharp edges for cutting the food. Next come the *canine* teeth with a sharp point for tearing it, while at the back are the *molars* with a broad flat top for grinding.

There are two sets of teeth: 1. the *temporary* or *milk teeth*, twenty in number—four incisors, two canines, and four molars in each jaw—which appear at from six months to two years, and 2. the *permanent teeth*, thirty-two in number—four incisors, two canines, known as *eye teeth* in the upper jaw and as *stomach teeth* in the lower jaw, four bicuspids, so called because they have two cusps where the molars have four or five, and six molars in each jaw—which come from the sixth to the twenty-first years. The first to appear are the two lower middle incisors, which come at the age of six months. The last to appear are the *wisdom teeth*, the farthest back of the molars, which come at the age of twenty-one years or thereabouts.

Each tooth consists of a crown or body above the gum, a neck, and a fang or root within the gum. The body is of dentine or ivory with a thin crust of enamel and contains the *pulp*, a vascular connective tissue containing many nerves. Beginning at the neck and covering the fang is a layer of cement or true bone.

The Sense of Taste.—The sense of taste lies chiefly in the *taste buds* as they are called which are filled with *gustatory cells* and are found in the papillæ of the tongue, principally in the *circumvallate papillæ* at the back of the tongue, which are few in number and arranged in a V-shape. There is also a certain power of taste in the tip and sides of the tongue but little in the upper surface or dorsum. Only five special tastes can be distinguished: bitter, sweet, acid, sour, and salt, but sometimes more than one can be distinguished at a time, as bitter and sweet. Every one can distinguish between different tastes but the power varies in different people and with different conditions. Certain tastes seem to be better distinguished in certain places, as sweet at the tip and bitter at the back of the tongue. Moreover, the sense of taste is very dependent upon the sense of smell, especially in the case of aromatic and savory substances, which one really does not taste but smell. If one held his nose and closed his eyes he would not know from the taste whether he was eating onion or apple. This leads to the habit of pinching the nose when taking nauseous medicines.

To be tasted a substance must be in solution. Friction against the tongue, lips or cheek increase the sense of taste. A temperature of 100° Fahrenheit favors taste, while both great heat and great cold impair it.

There are probably at least two nerves of taste, the lingual branch of the trifacial or fifth cranial and the gustatory branch of the glosso-pharyngeal.

Along with the sense of taste there are other senses in the mouth which play an important part, such as *pressure* and the *sense of heat and cold*, and it is often hard to distinguish them from the pure sensation of taste, which indeed is always accompanied by them.

Salivary Glands.—On either side of the mouth are three racemose glands for the secretion of the *saliva*, which serves to soften and lubricate the food and partially to digest starches by means of its ferment, *ptyalin*. The <u>*parotid gland*</u> is the largest and is below and in front of the ear, opening by *Stensen's duct*. The <u>*submaxillary gland*</u> is below the jaw toward the back on either side and its duct is *Wharton's duct*. The <u>*sublingual gland*</u> lies beneath the mucous membrane of the floor of the mouth and opens by eight to twenty tiny ducts beside the frenum, the *ducts of Rivinus*. The activity of the glands depends upon the blood supply; the more blood the greater their activity.

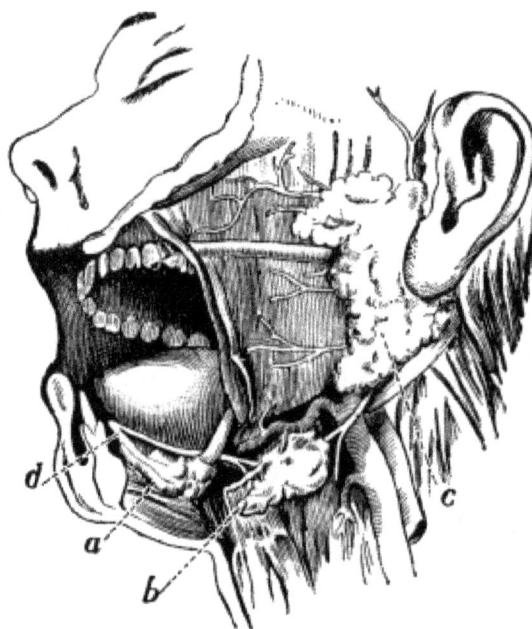

FIG. 20.—Dissection of the side of the face, showing the salivary glands: *a*, Sublingual gland; *b*, submaxillary gland, with its duct opening on the floor of the mouth beneath the tongue at *d*; *c*, parotid gland and its duct, which opens on the inner side of the cheek. (After Yeo.)

The Tonsils.—The tonsils vary in size and in *tonsillitis* swell and may even meet in the median line. They are frequently removed. When they are enlarged one often gets a *third tonsil* or *adenoids*, a lymphoid growth at the back of the pharynx which causes mouth-breathing by day and snoring by night. A child with adenoids is starved for air and what air is breathed in is not warmed. The growth should be removed.

A short frenum produces *tongue-tie*, which may be remedied by snipping. Cancer of the tongue is fairly common and necessitates a radical operation. In mumps the parotid glands are inflamed and enlarged.

The Ear.—The special organ of hearing is the ear, to which there are three parts, the external, the middle, and the internal ear.

The *external ear* consists of the *pinna* or expanded cartilaginous portion, for the concentration and direction of sound waves, and the *external auditory canal*, partly cartilage, partly bone, which is directed forward, inward, and downward and conveys sound to the middle ear.

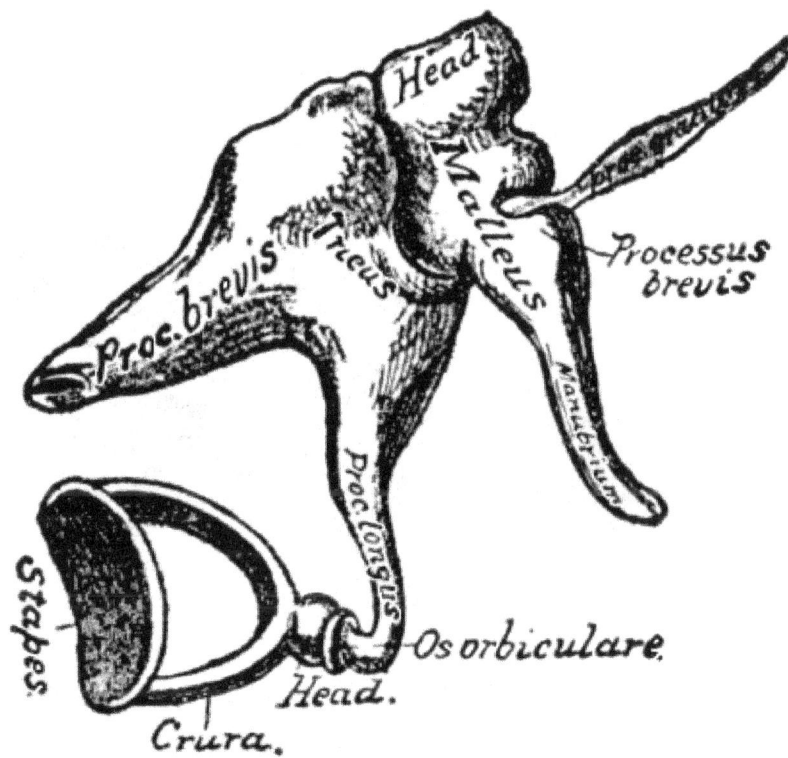

FIG. 21.—The small bones of the ear; external view (enlarged).
(After Gray.)

The *middle ear* or *tympanum* is an irregular cavity in the petrous portion of the temporal bone. Its outer wall is formed by the *membrana tympani* or *drum*, an oval translucent membrane placed obliquely at the bottom of the external auditory canal. The middle ear communicates with the inner ear through the *fenestra ovalis* or oval window and contains the *ossicles*, the *malleus* or hammer, the *incus* or anvil, and the *stapes* or stirrup, which are arranged in a movable chain from the drum to the oval window. The *malleus*, which is connected with the membrana tympani, articulates by its head with the body of the incus, while the stapes articulates with the *incus* by its head and is connected by its base with the margin of the oval window. Connection is made between the middle ear and the pharynx and the pressure of the air upon the drum made equal on either side by means of the Eustachian tubes. These tubes are about an inch and a half long, have cilia, and convey wax and other matter from the ear to the pharynx. Occasionally in a cold or for some other reason they become stopped up and trouble results in the middle ear. Some of the *mastoid cells* also connect with the middle ear and may become infected, causing mastoid disease.

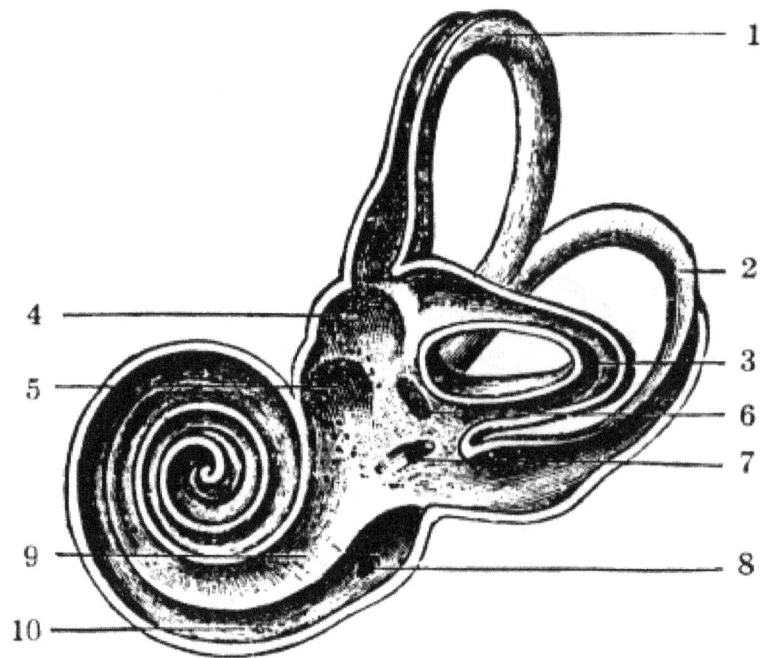

FIG. 22.—Interior view of left bony labyrinth after removal of the superior and external walls: 1, 2, 3, the superior, posterior, and external or horizontal semicircular canals; 4, fovea hemi-elliptica; 5, fovea hemispherica; 6, common opening of the superior and posterior semicircular canals; 7, opening of the aqueduct of the vestibule; 8, opening of the aqueduct of the cochlea; 9, the scala vestibuli; 10, scala tympani; the lamina spiralis separating 9 and 10. (From Quain, after Sömmerring.)

The *internal ear* consists of various chambers hollowed out in the petrous portion of the temporal bone. There is an *osseous labyrinth*, consisting of a central cavity known as the *vestibule*, three semicircular canals, and the cochlea and within the osseous labyrinth, surrounded by *perilymph*, is the *membranous labyrinth*, of like form, filled with the *endolymph*. Communication exists externally with the middle ear by the *round* and *oval windows* and internally with the *internal auditory canal*, through which passes the eighth cranial or auditory nerve, the special nerve of hearing, which is distributed to the inner ear only. When the auditory nerve enters the ear through this internal auditory meatus it divides into two branches, of which one goes to the vestibule and the other to the *organ of Corti*, a group of specially modified epithelial cells in the cochlea of the membranous labyrinth, which is very important in transmitting the impulses to the brain. The nerve also breaks up into very small branches and is distributed practically throughout the wall of the labyrinth.

The *sensation of hearing* is the result of impulses transmitted to the auditory nerve and so conveyed to the auditory center in the brain. It is caused by sound waves which travel through the air from their point of origin and enter the external ear. This collects and selects the waves of sound and helps one to a certain extent to determine the direction from which the sound comes. As they pass through the external meatus the sound waves are collected into a comparatively small area for transmission to the middle ear, where, by means of the drum, they set in vibration the chain of ossicles. Through these the vibrations are in turn transmitted to the oval window, being intensified in the process. Here again they are taken up by the perilymph, from which they pass through the wall of the membranous labyrinth to the endolymph, affecting the epithelial lining of the labyrinth in such a way that the impulses are transmitted to the auditory nerve, more particularly in the vestibule, from which the vibrations enter the cochlea. They also affect the cells of the organ of Corti in like manner as they pass from the perilymph to the endolymph. The membrane that covers the fenestra rotunda or round window relaxes and expands as the vibrations strike it, thus serving to eliminate the shock of impact.

Musical sounds are caused by rhythmical or regularly repeated vibrations, while irregular vibrations give rise to noises. In musical sounds loudness is determined by the height or amplitude of the vibrations, pitch by the length of the wave, and quality by the number of so called partial tones. A sensation of sound cannot be produced by less than 30 vibrations a second and the ordinary person cannot hear more than 16,000 vibrations a second. Different sounds can be distinguished when they follow each other as closely as by one one-hundredth of a second.

All sound does not come through the canal of the ear. The bones of the head vibrate and carry sound. So there are instruments for the deaf which are put in the ear and others which are placed between the teeth.

The _semicircular canals_ are not essential to hearing but have something to do with a person's power of maintaining his equilibrium. Injury to them may cause dizziness and loss of equilibrium.

The Eye.—One more feature, perhaps the most expressive, remains to be described, the eye. The senses are all modifications of the original cutaneous sensibility and the nerve of sight is no more sensitive to light than any other nerve. It therefore needs an end organ that is sensitive to the motions of the ether in order to give impressions of light. This organ is provided in the eye, which is not only itself capable of being moved in every direction, but is placed in the most movable part of the body, the head, which can be turned in almost a complete circle. The _eyeball_ is spherical and lies in the cavity of the orbit upon a cushion of fat, where it has a large range of sight but is securely protected from injury by its bony surroundings. The sunken eyes following protracted illness are due to the using by the system of the fat on which the eyeball ordinarily rests.

Each _orbital cavity_ is formed by the juncture of some seven bones and communicates with the cavity of the brain through the _optic foramen_ and through the sphenoidal fissure. Above the orbits are arched eminences of skin, the _eye-brows_, from which several rows of short hairs grow longitudinally and which serve to protect the eyes and to limit the amount of light to a certain extent, as in frowning.

Still further protection is afforded by the _eyelids_, longitudinal folds of skin, the one above, the other below, which close like curtains over the eye. Beneath the external layer of skin in the lids is fatty tissue and then the orbicularis palpebrarum muscle by means of which they are closed. They are kept in shape by the tarsal plates or cartilages, in whose ocular surface are embedded the _Meibomian glands_, whose secretion prevents the free edges of the lids from sticking together. Along these edges grows a double or triple row of stiff hairs, the _eye-lashes_, which curve outward so as not to interfere with each other and also to prevent the entrance into the eye of foreign bodies. Lining the inner surface of the lids and reflected thence over the anterior surface of the sclerotic coat of the eye is a mucous membrane, the *conjunctiva,* which is thick, opaque, and vascular on the lids but thin and transparent on the eyeball. The angles between the lids are known as the *internal* and the *external canthus.*

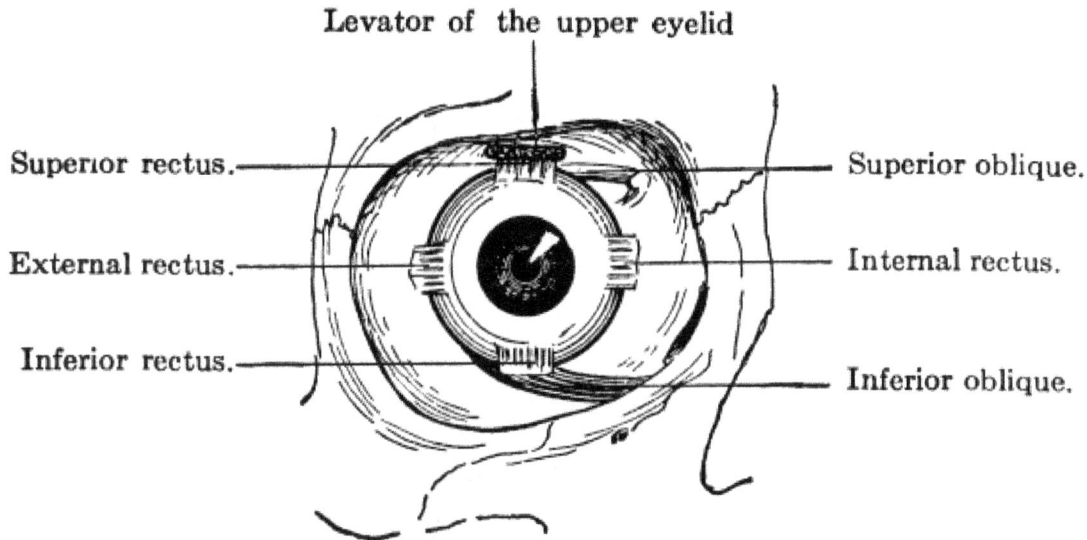

Levator of the upper eyelid

Superior rectus.

External rectus.

Inferior rectus.

Superior oblique.

Internal rectus.

Inferior oblique.

FIG. 23.—The external ocular muscles. (Pyle.)

Muscles and *Nerves.*—The eyeball is held in position by the ocular muscles, the conjunctiva, and the lids, while surrounding it, yet allowing free movement, is a thin membranous sac, the *tunica vaginalis oculi.* The superior and inferior recti muscles at the upper and lower edges of the ball turn the eye up and down; the internal and external recti at the inner and outer edges turn the eye inward and outward; and the superior and inferior oblique rotate the eye. The nerves supplying these muscles are the third or motor oculi, the fourth and the sixth.

The *lachrymal gland,* which is about the size and shape of an almond, is situated at the upper and outer part of the orbit. It secretes a fluid which keeps the anterior surface of the eye bathed in moisture and is ordinarily drained away through the lachrymal sac in the inner canthus, whence it passes by the lachrymal ducts into the nose. When the amount secreted is excessive, it overflows the lower lid as *tears.*

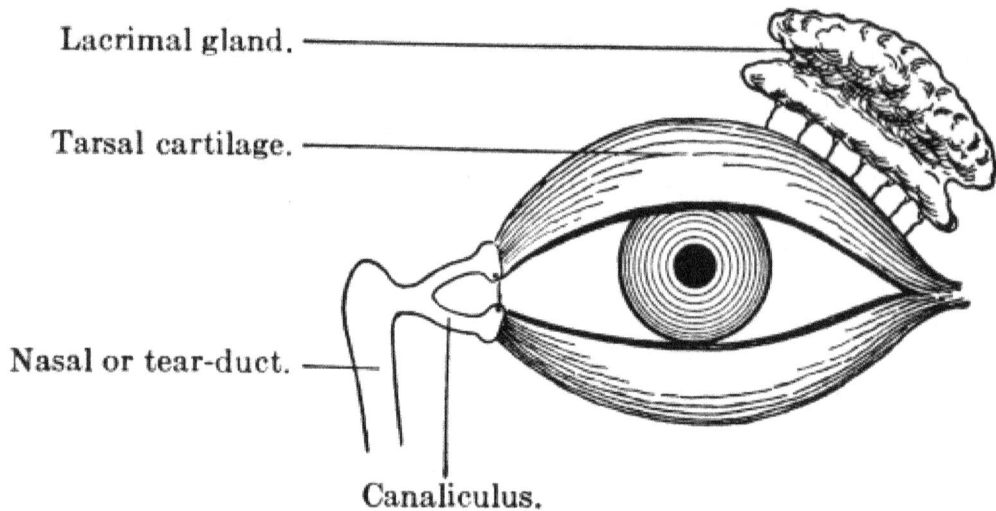

Lacrimal gland.

Tarsal cartilage.

Nasal or tear-duct.

Canaliculus.

Fig. 24.—Diagram of the lacrimal apparatus. (Pyle.)

Coats of Eye.—The membranes or coats of the eye are three in number: an outer or sclerotic, a middle or vascular, and an inner or sensitive.

The *sclerotic coat* is a rather thick, fibrous, protective membrane. Where it passes in front of the iris, however, it is thinner and transparent and is known as the cornea. The cornea projects somewhat and, as it were, resembles a segment of a smaller sphere set into the rest of the sclerotic.

The middle or vascular coat, known as the choroid, carries blood-vessels for the retina or sensitive coat in its inner layer and has an outer layer of pigment cells that excludes light and darkens the inner chamber of the eye. The folds of the choroid at its anterior margin contain the ciliary muscles and are known as the *ciliary processes*, while the name iris is given to the little round pigmented, perforated, curtain-like muscle just in front of the crystalline lens. The posterior surface of the iris is covered with a thick layer of pigment cells to prevent the entrance of light except through the central opening or *pupil*, and its anterior surface also has pigment cells that give it its color, though the difference in the color of people's eyes is due rather to the amount of pigment present than to its color, a small amount of pigment being present in blue eyes and a large amount in brown and black eyes. Variations in the size of the pupil are brought about by contractions of the circular and radiating fibers of the iris, contraction of the circular fibers making it smaller and those of the radiating larger. The pupil is constricted for near objects and during sleep, and is dilated for distant objects. In a dull light also it dilates to let in more light, and in a bright light it contracts. The appearance of the pupil is often important as a means of diagnosis and in etherization.

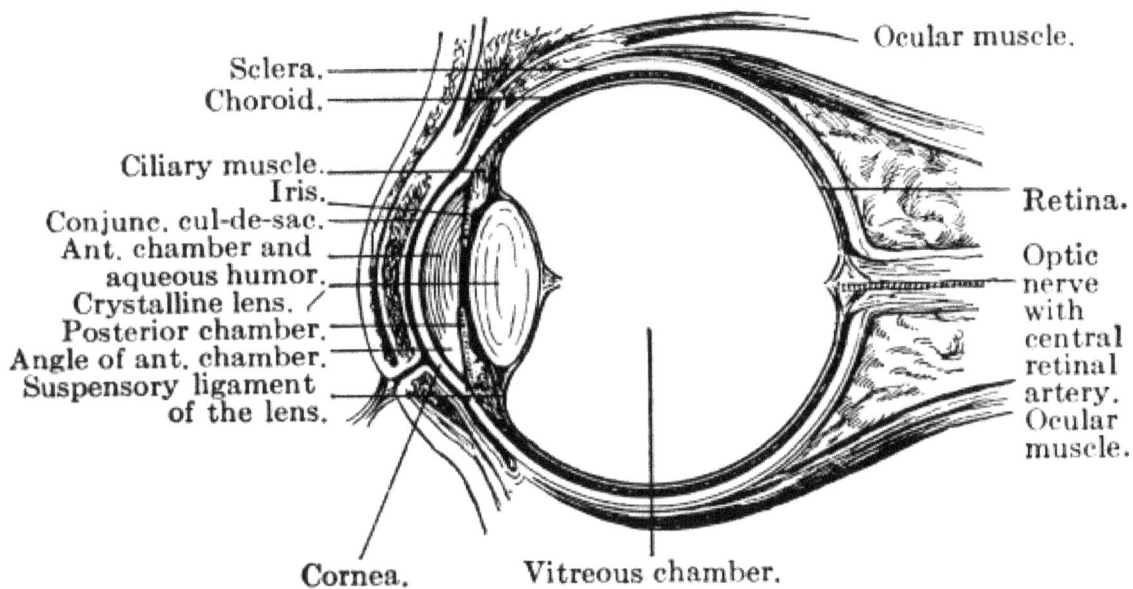

FIG. 25.—Vertical section through the eyeball and eyelids. (Pyle.)

Lastly there is the innermost *sensitive coat* or *retina*, which has eight layers, the outer one containing some pigment cells and the next the rods and cones, in which the power of perception is supposed to lie, branches of the optic nerve being distributed over it in all directions. In fact, the retina is formed by a membranous expansion of the optic or second cranial nerve, the special nerve of sight, which passes into the orbit through the optic foramen at the back and enters the eyeball close to the *macula lutea* or *yellow spot*. The exact spot where the optic nerve enters the retina is not sensitive and is known as the *blind spot*. In the center of the macula lutea, however, which is in the middle of the retina posteriorly, is a tiny pit, the *fovea centralis*, in which all the layers of the retina except the rods and cones are absent, and at this point vision is most perfect. It is, therefore, always turned toward the object looked at, and when one wishes to see an object distinctly, he must keep moving his eyes over it that the rays from each part may fall in turn upon the fovea centralis.

Directly behind the pupil is the *crystalline lens*, a rather firm gelatinous body enclosed in a capsule, which is transparent in life but opaque in death. The lens is doubly convex and is held in place by the suspensory ligaments, which arise from the ciliary processes. In front of it is the anterior chamber of the eye, filled with a thin watery fluid called the _aqueous humor_, while the larger space back of it, occupying about four-fifths of the entire globe, is filled with a jelly-like substance known as the _vitreous humor_.

The chief artery of the eye is the _ophthalmic_.

Light Rays.—The eye is practically a camera and its principal function is to reflect images. Although there are several refracting surfaces and media, for practical purposes the cornea alone need be considered. Except for those rays which enter the eye perpendicularly to the cornea, whose line of entrance is called the _optic axis_, all rays are refracted when they enter the eye and the point at which they meet and cross each other behind the cornea is called the _principal focus of the eye_. To focus properly, all the rays from any one point on an object must meet again in a common point upon the retina, their _conjugate focus_. In the normal eye all the rays from an object are focused on the retina and form upon it an image of the object which, as in the camera, is inverted, because of the crossing of the rays behind the cornea. Once focused on the retina the light traverses the various layers to the layer of rods and cones, where chemical action takes place and affects the little filaments of the optic nerve, by which the message is carried to the brain.

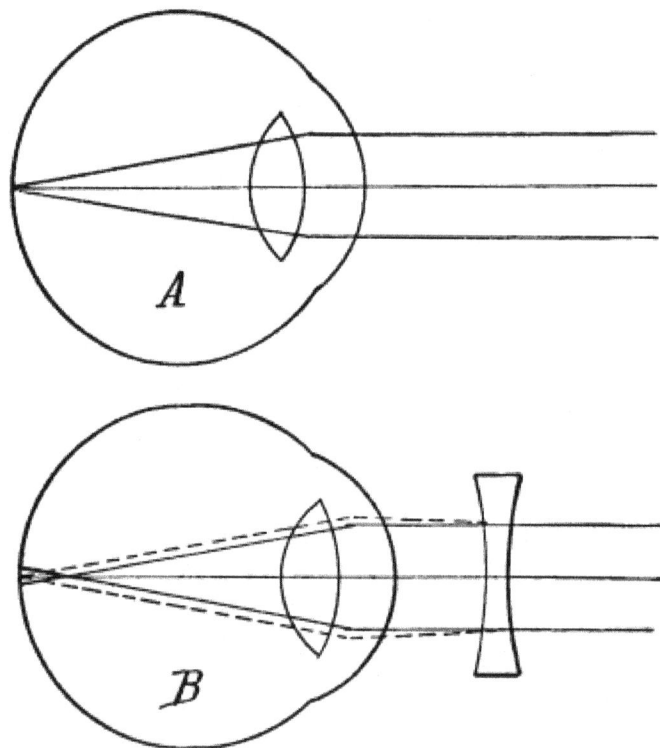

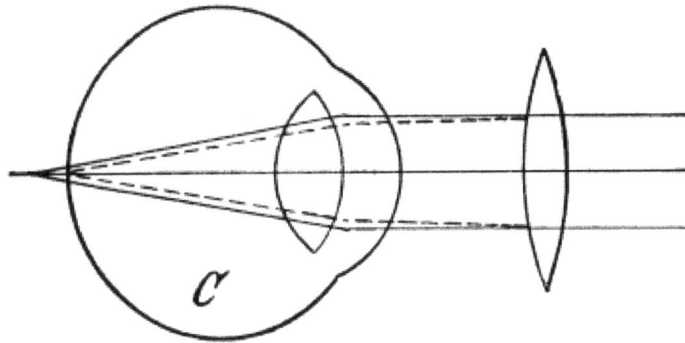

FIG. 26.—Diagram showing the difference between (*A*) emmetropic, (*B*) myopic and (*C*) hypermetropic eyes. (American Textbook of Physiology.)

When the eye is at rest the pupil and lens are in their normal condition and at such times the eye sees only distant objects. The ability of the eye to focus upon objects at different distances is called *accommodation* and to accomplish it three things are necessary: 1. change in the shape of the lens; 2. convergence of the axes of the eyes, and 3. narrowing of the pupils.

When the eye is directed toward distant objects, the muscle fibers in the ciliary processes relax, causing tightening of the suspensory ligaments and consequent flattening of the surface of the lens. Otherwise an image would be formed in front of the retina; for the greater the convexity of the lens, the greater the angle of refraction. Such accommodation is passive and so not fatiguing. To look at nearby objects, on the contrary, the ciliary muscles contract, drawing the choroid forward and allowing the suspensory ligaments to relax, so that the lens bulges in front. This is an exertion.

In order to accommodate properly, moreover, both eyes must work together and the axes of both eyes must be directed toward the object. Therefore, in looking at nearby objects the axes of the eyes converge, drawn by the internal recti muscles. In *strabismus* or *cross eye*, where the axes of both eyes cannot be directed toward the object at the same time, the rays fall upon one part of one eye and upon a different part of the other eye and two separate images are seen.

Finally there is concentric narrowing of the pupil by contraction of the circular fibers of the iris, by which means various side rays that would come to a focus outside the retina are excluded.

All the muscles of accommodation, the ciliary muscles, the internal recti, and the sphincter pupillæ, are under the control of the third nerve.

Connected with this power of accommodation and dependent on it are the two conditions of *near-sightedness* or *myopia* and *far-sightedness* or *hypermetropia*.

The normal eye is *emmetropic* and is almost perfectly spherical, but in the near-sighted or myopic eye the ball, instead of being round, is flattened from above down and so bulges in front. Consequently, owing to the greater distance from the lens to the retina, images are formed in front of the retina. Only nearby objects can be seen clearly, because the farther the object from the eye the farther in front of the retina the image is formed. Concave glasses are worn to enable near-sighted people to see at a distance. Hypermetropic or far-sighted eyes are flattened from before backward and can see only objects at a distance clearly, as those nearby form images behind the retina. For such eyes convex glasses are worn.

As the ordinary person approaches middle life, he becomes able to see better at a distance than near to. This *presbyopia,* as it is called, which is practically far-sightedness, is due to a partial loss of the power of accommodation in the lens, the result of a general loss of elasticity in the parts.

Another very common defect is *astigmatism*, a failure of the rays to focus upon a point, owing generally to a flattening in the surface of the cornea.

Color perception is also an important function of the eye. The waves of hyperluminous ether when of a certain rate of vibration give the sensation of heat and when their vibrations are more rapid they give the sensation of light. Each of the primary colors of the spectrum gives off a pretty definite number of light rays which travel through the air and enter the eye, the number of rays determining the color thrown upon the retina and the velocity determining the intensity of the color. Occasionally when light is passing through into the eye it is broken up as in a prism and the person gets a sensation as of all sorts of colors, *chromatic aberration*. Total or partial absence of sensitiveness to color is called *color blindness*. It is commonest in the form of inability to distinguish between red and green and is probably due to a defect in the retina.

Sometimes a hair follicle on the lid becomes infected and a *sty* is formed. *Pink eye* is conjunctivitis or inflammation of the conjunctiva. A Meibomian duct may become stopped and cause bulging, or there may be a sagging down or *ptosis of the upper lid* in certain diseases, as meningitis, apoplexy, and more especially syphilis. *Rodent ulcer* often begins by the eye or on the cheek.

CHAPTER V.
THE NERVOUS SYSTEM.

The nervous system, which regulates all the vital processes of the body, physical and chemical, and which is situated partly in the head and partly in the trunk, may well form the connecting link between the description of the head and that of the trunk. It has two divisions, the cerebro-spinal system and the sympathetic system. The former consists of the cerebrum or brain proper, the cerebellum or little brain, the pons Varolii, the medulla oblongata, the spinal cord, and the cranial and spinal nerves; the latter of a series of ganglia or aggregations of nerve centers. The brain, which includes the cerebrum, cerebellum, pons, and medulla, occupies the cranium and the spinal cord is contained within the bony framework of the spinal column. In the male the brain weighs about 49 ounces and in the female 44, while in an idiot it seldom weighs more than 23 ounces.

The **cerebrum** or brain proper has two parts or hemispheres, roughly oval in shape, each of which has five lobes separated by fissures, the frontal, parietal, occipital, and temporo-sphenoidal lobes, and the central lobe or *island of Reil* at the base of the brain. The chief fissures are the *longitudinal fissure*, the *fissure of Sylvius* at the base of the brain, and the *fissure of Rolando* between the frontal and parietal lobes. There are also five serous cavities called *ventricles*, the two lateral and the third, fourth, and fifth ventricles, of which the first two, one in either hemisphere, are the most important. Around these cavities is the *brain substance*, which is made up of two tissues, the white and the gray, the latter forming the outer part of the brain to the depth of perhaps half an inch, and the white matter forming the rest. The outer or gray part is called the *cortex* and is largely made up of nerve cells. It might be called the active part of the brain. The white part consists largely of nerve fibers which are given off from the nerve cells and are carried down into the spinal cord.

The surface of the brain is convoluted, the ridges being separated by deep furrows or sulci, by which means a great extent of gray matter is secured. The furrows contain fluid from the subarachnoid spaces and vary in number and depth according to intelligence. While the convolutions are not uniform in all brains, the principal ones are constant.

Both the brain and the spinal cord are covered by three membranes, the dura mater, the arachnoid, and the pia mater. The *dura mater* is dense and fibrous and lines the interior of the skull, being firmly adherent to it at many points. In fact, it constitutes the internal periosteum of the cranial bones. The *arachnoid* is a delicate serous membrane, with two layers, lubricated to prevent friction, which divides the space between the dura mater and the pia mater, bridging over the convolutions and enclosing the subdural and subarachnoid spaces which are connected with lymphatics and contain a serous secretion, the *cerebro-spinal fluid*. This fluid forms an elastic water cushion, on which the brain rests, and prevents concussion. The *pia mater* is vascular, containing blood-vessels, lymphatics, and nerves, and is closely attached to the surface of the brain, dipping down into all the sulci.

At the base or under surface of the brain are some very important structures. The *olfactory bulbs* lie beneath the frontal lobe and projecting back is the *olfactory tract*, through which the olfactory nerves come from the brain. Back of the olfactory tract is the *optic commissure* where the optic nerves coming from the brain cross each other. And back of the commissure again is the *optic tract*, where the optic nerves emerge from the brain. At the base of the brain are also the exits of the twelve cranial nerves.

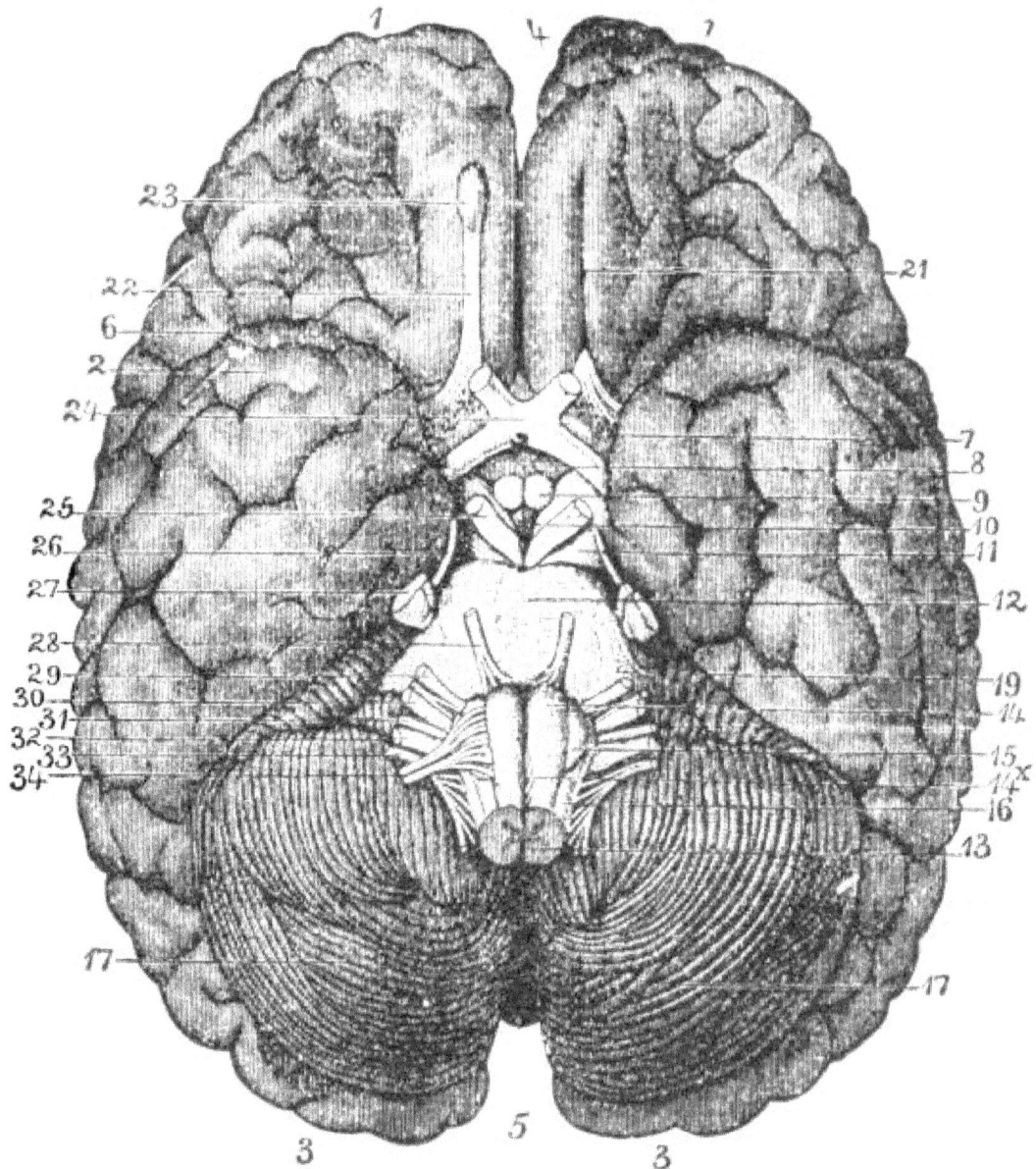

FIG. 27.—Base of brain. (Leidy.) 1, 2, 3, cerebrum; 4 and 5, longitudinal fissure; 6, fissure of Sylvius; 7, anterior perforated spaces; 8, infundibulum; 9, corpora albicantia; 10, posterior perforated space; 11, crura cerebri; 12, pons Varolii; 13, junction of spinal cord and medulla oblongata; 14, anterior pyramid; 14^x, decussation of anterior pyramid; 15, olivary body; 16, restiform body; 17, cerebellum; 19, crura cerebelli; 21, olfactory sulcus; 22, olfactory tract; 23, olfactory bulbs; 24, optic commissure; 25, motor oculi nerve; 26, patheticus nerve; 27, trigeminus nerve; 28, abducens nerve; 29, facial nerve; 30, auditory nerve; 31, glosso-pharyngeal nerve; 32, pneumogastric nerve; 33, spinal accessory nerve; 34, hypoglossal nerve.

Upon entering the brain the *arteries* run a tortuous course, the tortuosity breaking the force of the blood stream in the small vessels where congestion would be with difficulty relieved. The basilar artery, which is formed by the juncture of the two vertebrals, divides into the two posterior cerebrals, each of which joins one of the anterior cerebrals by a posterior communicating artery. The two anterior cerebrals also are joined by an anterior communicating artery, thus completing the circle. The circle thus formed at the base of the brain is called the *circle of Willis* and provides for a good supply of blood in event of an accident to any vessel. The blood is returned to the general circulation through the cerebral veins and sinuses formed by the separation of the dura mater into two layers.

The **cerebellum** is about one-seventh the size of the cerebrum and weighs about 5 ounces. It lies in the lower occipital fossæ of the skull and is oblong in shape and divided into two lateral hemispheres by a transverse fissure. It is made up of both white and gray matter, of which the former predominates, the gray being external as in the cerebrum. The cells are about the same as in the cortex and its surface is traversed by queer furrows. Of its function little is known but it probably plays a most important part in the coördination of the nervous and muscular acts by which the movements of the body are carried on.

At the back of the cerebrum and below the cerebellum is the **pons Varolii**, which forms a connecting link with the medulla oblongata or bulging part of the cord. It is made up essentially of white matter or nerve fibers, though there is a small amount of gray matter in which are found the nuclei of some of the cranial nerves.

In the **medulla oblongata**, which is about 1 inch long and extends from the pons Varolii to the upper border of the atlas or first cervical vertebra, the gray matter is not necessarily external to the white but is found in patches in the white. The gray matter here corresponds more or less to that of the spinal cord and the white matter is continuous with that of the cord. From the medulla arise the fifth to twelfth cranial nerves and the vasomotor nerves. The cardiac nerve has its center here and here too are the centers of respiration, phonation, deglutition, mastication, and expression. In the medulla the nerves that arise in the cerebrum cross over from one side of the body to the other on the crossed pyramidal tracts. The importance of this crossing of the nerve fibers is seen in apoplexy, when a blood-vessel is ruptured in the brain and hemorrhage causes pressure, generally on the motor tract. Paralysis of the nerves and of the muscles to which they go results. The paralysis is generally of one side of the body, the opposite side from that on which the injury occurred. The seat of injury in the brain or cord can frequently be determined by the situation and extent of the paralysis.

Spinal Cord.—Extending down from the medulla through the spinal column is the cord. Its length from the foramen magnum, where it begins, down through the vertebræ to the lower border of the first lumbar vertebra, where it ends in a very fine thread-like process with no special function, called the *filum terminale*, is 17 to 18 inches. Just before it ends a number of nerves are given off in a tail-like expansion known as the *cauda equina* or horse's tail. It is not uniform throughout its length but presents two enlargements, a cervical enlargement in the lower cervical region, and a lumbar enlargement in the lower dorsal region, where the nerves are given off to the arms and legs respectively. The membranes are the same as those of the brain and are continuous with them, but here the dura mater is not attached to the bony walls enclosing it. For the cord does not fit closely into the canal but is as it were suspended in it. The *subarachnoid space* communicates with the ventricles of the brain by the *foramen of Majendie* and is filled with *cerebro-spinal fluid* for the protection of the cord. In cerebro-spinal meningitis or spotted fever this fluid is infected and for diagnosis lumbar puncture is performed.

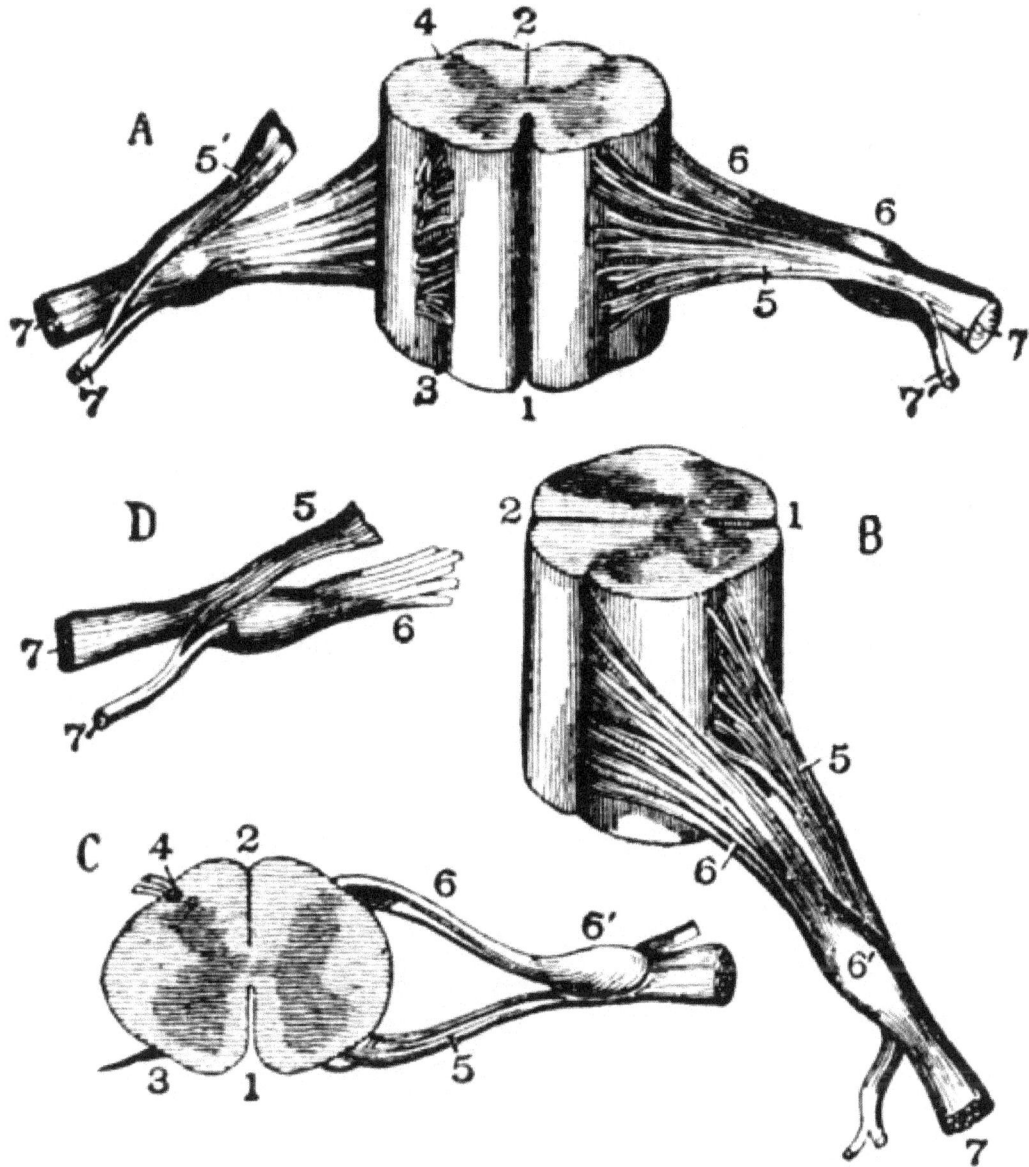

Fig. 28.—Different views of a portion of the spinal cord from the cervical region, with the roots of the nerves. In *A* the anterior surface of the specimen is shown, the anterior nerve root of its right side being divided; in *B* a view of the right side is given; in *C* the upper surface is shown; in *D* the nerve roots and ganglion are shown from below: 1, the anterior median fissure; 2, posterior median fissure; 3, anterior lateral depression, over which the anterior nerve roots are seen to spread; 4, posterior lateral groove, into which the posterior roots are seen to sink; 5, anterior roots passing the ganglion; 5′, in *A*, the anterior root divided; 6, the posterior roots, the fibers of which pass into the ganglion, 6; 7, the united or compound nerve; 7′, the posterior primary branch seen in *A* and *D* to be derived in part from the anterior and in part from the posterior root. (Allen Thomson.)

If a cross-section of the cord is made, it is found to have a pretty definite structure. It is roughly circular and is divided by certain fissures, of which the most important are the anterior and posterior median, the latter being rather a dividing line or septum. By them it is divided into halves connected by a small band in the middle called the *commissure*. The white matter is exterior to the gray and is divided by it into four columns, which again are divided into tracts according to certain groups of nerves that travel through them. The most important tract is the direct pyramidal tract in the anterior column. The gray matter is arranged in the form of a letter H practically, consisting of two lateral halves, more or less crescentic in outline, connected by a narrow band, the *gray commissure*. Each half is divided into two *horns*, the anterior, toward the front of the cord, and the posterior, toward the back, the former being generally much thicker and heavier than the latter. The structure of the gray and of the white matter is essentially the same as in the brain, but the proportion varies in different parts of the cord, the white predominating in the cervical region and the gray being much better developed in the lumbar region, where the nerve cells for control of the lower extremities occur. The gray is least well developed in the dorsal region. Through the center of the cord runs a small hole or canal filled with cerebro-spinal fluid, the *central canal of the cord*.

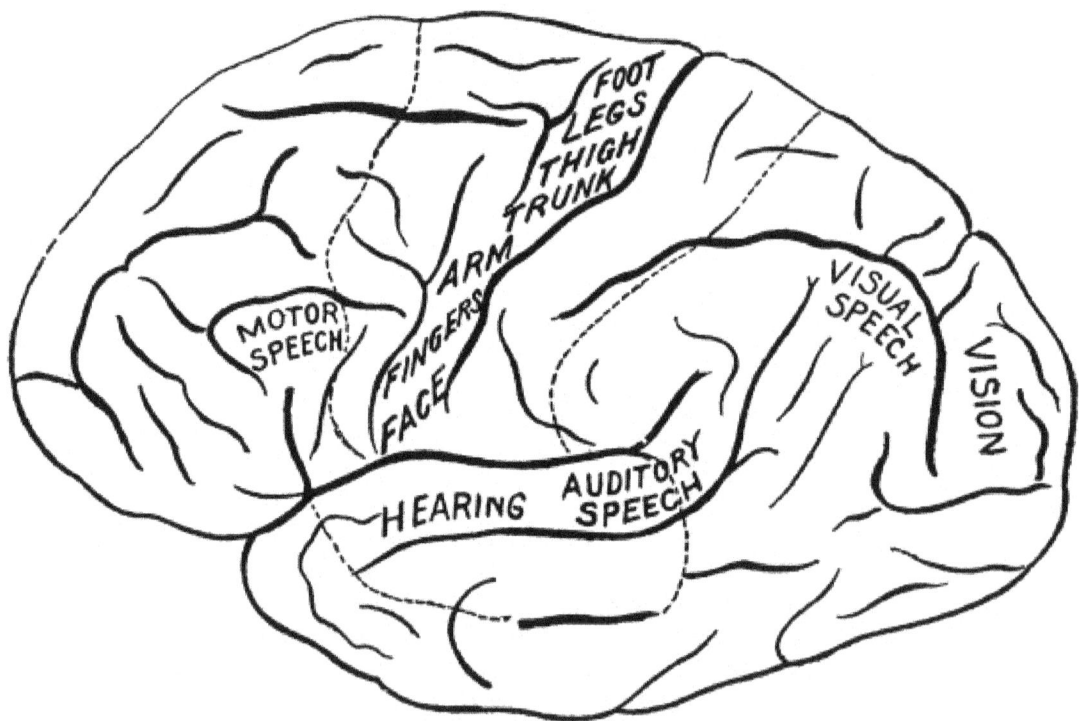

The brain is the seat of intelligence and will, the center of all voluntary action. Molecular change in some part of the cerebral substance is the indispensable accompaniment of every phenomenon of consciousness. Indeed, the brain is never in a state of complete repose, there being dreams even during sleep. The brain is not sensitive to injury in the sense of pain. It can be lacerated without much pain.

Various centers exist in the brain, of which the most important perhaps is the *motor center*. The *visual center* is in the occipital lobe, the *auditory center* in the temporal lobe, the *speech center* in the third left frontal convolution. Thus the impulses of the senses have been located, though the function of many parts, the so called *silent areas*, are still in obscurity.

The *motor center*, that is, the center for motion of the skeletal muscles, is situated about the fissure of Rolando and is divided into three parts, one for the legs, one for the face, and one for the arms, the one for the legs being uppermost and the others below in the order mentioned. Fibers from these cells extend down through the brain and cord to the muscles, the fibers being collected into well-recognized bundles and the whole known as the *motor tract*. There may be one long fiber from a cell in the brain down through most of the cord or there may be a succession of shorter fibers that are not actually connected but are in close contact with each other. In the upper pons the fibers for the face cross to the opposite side, while the rest keep on down through the medulla, and as they emerge from the medulla they too cross to the other side and keep on down in the crossed pyramidal tract. A few fibers do not cross but come down the direct pyramidal tract, which, however, disappears part way down. The *crossed pyramidal tract* is the true motor tract and in it the fibers are continually sending branches to the cells in the gray matter, where they connect with the anterior horn.

The anatomy of the *sensory tract* is not so well understood. By it impulses are sent to the brain by the peripheral organs, practically the surface of the body. The sensory fibers connect with the sensory cells in the posterior horn, from which fibers are sent to the brain, practically the reverse of motor action. There are three chief sensory tracts, which are supposed to transmit different sensations, one pain, one muscular sensations, and the third sensations of touch. All these tracts, of which the chief is the *direct cerebellar tract*, in passing up the cord pass to the opposite side at different levels and then go on to the cortex of the brain.

The action of the nerves is similar to reflex action, only that an effort of will is needed to send an impulse from the brain. It is by the help of the brain along this line that an infinity of artificial reflexes or habits is acquired, for which volition is needed in the beginning but which are later done unconsciously. Herein lie the possibilities of all education.

The brain and spinal cord work together, the cord acting as a medium between the brain, in which all the higher psychical processes, such as will, thought, etc., originate, and the muscular apparatus. The cord, however, has some action entirely independent of the brain, as is seen in *reflex action*. This action is entirely involuntary, so that the cord is sometimes spoken of as the seat of involuntary action, commonly called reflex action. All unconscious acts are reflex acts, as when the hand is drawn away from a hot iron. If an impulse is sent along one of the sensory fibers, it enters the cord through the posterior horn, where its nerve cell is found. Then, through some connection between the nerve cell of the sensory fiber and that of the motor fiber the impulse is transmitted to the motor cell and another impulse is sent out of the cord along the motor fiber of the nerve to the muscle. One of the commonest reflexes is the *knee-jerk*. Reflex action is important because the reflexes are interfered with, delayed, destroyed, or increased in different diseases. The time normally required for a reflex act is very brief, that for the knee-jerk being about three one-hundredths of a second.

The nerves of the head, known as the *cranial nerves*, arise from the brain, while the rest of the body is supplied by the spinal nerves, which come off at intervals from the spinal cord. The cranial nerves consist of twelve pairs: (1) The olfactory or nerve of smell, (2) the optic or nerve of sight, (3) the motor oculi, (4) the patheticus, which controls the eye, (5) the trigeminus or trifacial, a nerve of general sensation, motion, and taste, (6) the abducens, a motor nerve, (7) the facial nerve of the face, ear, palate, and tongue, (8) the auditory or nerve of hearing, (9) the glosso-pharyngeal, nerve of sensation and taste, (10) the pneumogastric or vagus, which is both motor and sensory and governs respiration, the heart, and the stomach, (11) the spinal accessory, to the muscles of the soft palate, and (12) the hypoglossal, the motor nerve to the tongue.

The *spinal nerves* also are arranged in pairs: Eight cervical pairs, twelve dorsal or thoracic, five lumbar, five sacral, and one coccygeal, these titles denoting their point of origin near the vertebra of the same name. Each of these nerves arises by two roots, an anterior motor root from the anterior horn of gray matter and a posterior sensory root from the posterior horn, the latter having a ganglion upon it. After emerging from the cord the two roots unite to form the nerve, that the nerve may contain both motor and sensory fibers. The motor fibers are called efferent because they carry impulses *from* the cord, while the sensory are called afferent because they carry impulses back *to* the cord. After leaving the cord the nerves unite to form plexuses, which again divide into various nerve trunks and are distributed to the muscles.

The first *cervical nerves* pass out of the spinal column above the first cervical vertebra and the other cervical nerves below that and the succeeding vertebræ, while the other spinal nerves emerge each below the corresponding vertebra, as the first dorsal below the first dorsal vertebra, etc. After emerging they break up into a large anterior division and a small posterior division, the posterior branches supplying the spine and the dorsal muscles and skin, the anterior the rest of the trunk and the limbs. The *cervical plexus* is formed by the anterior divisions of the first four cervical nerves, the *brachial plexus* by the last four cervical and the first dorsal or thoracic nerves, the *lumbar plexus* by the four upper lumbar, and the *sacral plexus* by the last lumbar and the four upper sacral nerves.

The only important branch of any of the four upper cervical nerves, which in general supply the neck and shoulders, is the *phrenic*, which is distributed to the pericardium, the pleuræ, and the under surface of the diaphragm.

The *brachial plexus*, as its name implies, supplies the arms and has a number of important branches, as the circumflex to the shoulder, the musculo-cutaneous to the upper arm, the elbow-joint, and the outer surface of the forearm, the internal cutaneous to the inner side of the arm, the median to the pronators and flexors and the fingers on the radial side, and the ulnar to the elbow and wrist-joint. The musculo-spiral runs down the spiral groove to the external condyle of the humerus or upper arm bone, where it divides into the radial and the posterior interosseous, the former going to the thumb and two adjacent fingers and the latter to the wrist-joint and the muscles on the back of the forearm. Sometimes, in fracture of the humerus the callus thrown out pinches the musculo-spiral and causes pain.

The dorsal or *thoracic nerves* supply the back with their posterior divisions and their anterior divisions are the *intercostal nerves*.

The *lumbar nerves* supply the abdomen, pelvis, and thigh, the chief branches being the ilio-hypogastric to the abdomen and gluteal region, the ilio-inguinal to the inguinal region and scrotum, the external cutaneous and genito-crural to the thigh, and the obturator to the thigh and the hip and knee-joints. The anterior crural descends beneath Poupart's ligament and divides into an anterior and a posterior division which supply the thigh muscles, its branches going to the pelvis.

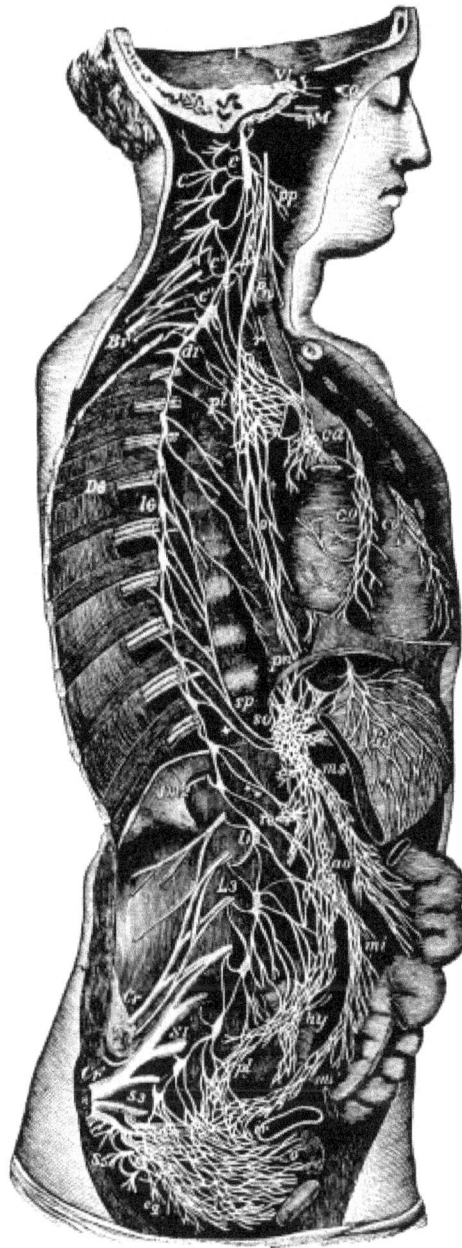

Fig. 30.—Diagrammatic view of the sympathetic cord of the right side, showing its connections with the principal cerebro-spinal nerves and the main preaortic plexuses. (Reduced from Quain's anatomy.)

The *sacral plexus* supplies the organs of the pelvis, the thigh, and the leg. Its chief branches are the great sciatic, the largest nerve in the body, and the small sciatic, which go to the buttocks and thigh. The great sciatic runs down the back of the thigh and divides at the lower third of the thigh into the internal and external popliteal nerves, the former of which passes along the back of the thigh to the knee, where it becomes the posterior tibial, which in turn divides at the ankle into the internal and external plantar. The external popliteal descends along the outer side of the popliteal space and divides an inch below the head of the fibula into the anterior tibial, which supplies the flexors and skin of the ankle-joint, and the musculo-cutaneous, which sends branches to the skin of the lower leg and the dorsum of the foot.

The Sympathetic System.—Joined to the cerebro-spinal system by intervening cords is the sympathetic system. This is made up of two series of ganglia, one on either side of the spinal column, connected by longitudinal bands and extending from the base of the skull to the coccyx. They do not form an independent nervous system, each ganglion, which seems to resemble the motor cells of the spinal cord, being connected by motor and sensory fibers with the cerebral system.

The _sympathetic nerves_ are mostly gray, non-medullated fibers and are distributed to viscera, secreting glands, and blood-vessels, whose movements are involuntary and feelings obtuse. They form networks upon the heart and other viscera and send branches to the cranium to the organs of special sense. There are three main plexuses: The _solar plexus_ behind the stomach, which supplies the abdominal viscera; the _hypogastric plexus_ in front of the prominence of the sacrum, whose nerves go to the pelvic organs; and the _cardiac plexus_ behind the aortic arch for the thoracic viscera.

Over these nerves one has no control. A blow in the region between the costal cartilages and below the sternum is a solar plexus blow and is very upsetting.

The sympathetic system serves to maintain vitality in all the important portions of the system and one of its important functions is to keep up communication between one part and another, so that when any organ is affected the others will act accordingly and help out to the best of their ability.

CHAPTER VI.
THE BACK.

Fig. 31.—
The spinal column.
(Church and Peterson.)

The Spine.—The trunk may be roughly divided into the back, the chest or thorax, the abdomen, and the pelvis. By the back is denoted the spinal column with its muscles, blood-vessels, etc., and the spinal cord already described. The spine or vertebral column, which serves the double purpose of holding the body erect and of protecting the cord, is usually about two feet, two inches in length. In its course there occur several curves, which serve to give springiness and strength and, with the intervertebral cartilages, to mitigate the force of concussion from blows and falls. The curve is convex forward in the cervical region, convex backward in the dorsal, forward in the lumbar, and backward again in the sacral region. There is most freedom of motion in the cervical region.

As is the case with the other bones, the *vertebræ* are specially adapted in shape and size to the needs they are called upon to fill. Strength and flexibility, with a minimum bulk, a channel for the cord, and passages for the numerous nerves and blood-vessels are some of the requirements which, in combination, they meet to an astonishing degree. They are thirty-three in all, and are divided into groups according to the region in which they occur: seven cervical in the neck, twelve dorsal or thoracic, five lumbar, five sacral, and four coccygeal.

Although the vertebræ of the different groups differ more or less in size and shape in accordance with the various demands of their positions, they all have certain general characteristics. Each has a body, two laminæ, two pedicles, two transverse processes, and one spinous process. The *pedicles* extend back from the body on either side and support two broad plates of bone, the *laminæ,* whose juncture at the back completes the *spinal foramen* for the passage of the cord. At their juncture is the *spinous process*, which can be felt beneath the skin, while the *transverse processes* project from the juncture of the laminæ with the pedicles. All the processes are for the attachment of muscles that move the spine. The *body* is formed of cancellous bone with a compact layer outside. Transversely it is slightly oval, while its upper and lower surfaces are flat, except in the cervical region, where the upper surface is concave laterally and the under convex laterally and concave from before back. Between the bodies are disks of fibro-cartilage, which increases motion and springiness. The *spinous process* or *spine* is short in the cervical region, long and directed downward in the dorsal region, thick and projecting almost straight out in the lumbar region. The pedicles are notched above and below so that when articulated the notches of two vertebræ join to form the *intervertebral foramen* for the outward passage of nerves and the inward passage of blood-vessels.

The distinguishing mark of the *cervical vertebræ* is the foramen in each transverse process, through which the vertebral arteries run to the skull. They are also smaller than the dorsal and lumbar vertebræ. The *dorsal vertebræ* are distinguished by having on the transverse processes and on the body smooth articular surfaces called facets and demi-facets for articulation with the ribs. The *lumbar vertebræ* are the largest and heaviest and have the thickest spine. By the time the sacral region is reached, however, the vertebræ have only a rudimentary spinous process. Moreover, in adult age the sacral bones grow together and form one triangular bone, the *sacrum*, which has a broad base called the *promontory of the sacrum* and a blunt apex. It is concave in front and convex behind and has an articulating surface for joining the pelvic bones. In the case of the *coccyx* also the four original bones, all rudimentary in character and supposed to be the survival of a tail, grow together to form one bone. Together the sacrum and coccyx form the posterior wall of the true pelvis.

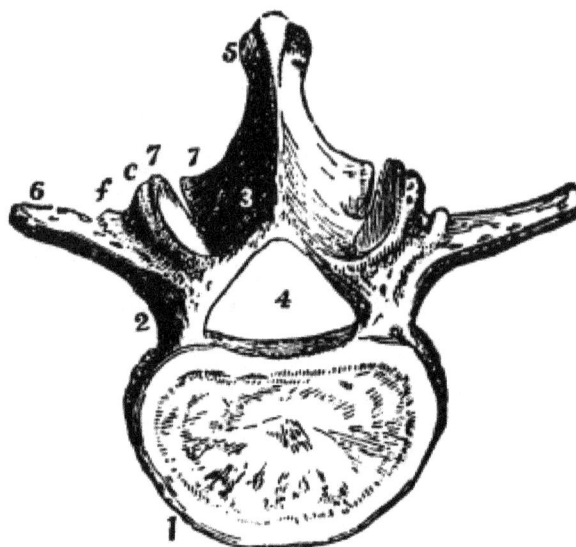

FIG. 32.—A type of vertebra. (Leidy.) 1, Body; 2, pedicle; 3, lamina; 4, spinal foramen; 5, spinous process; 6, transverse process; 7, articular process.

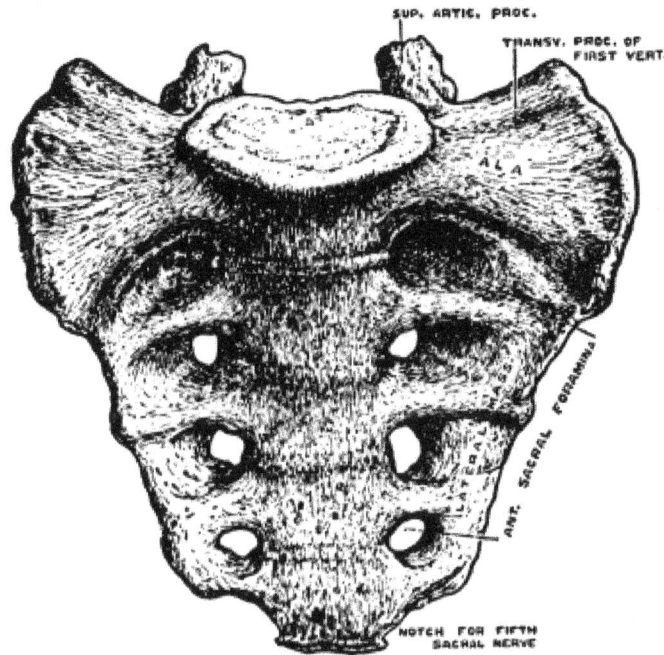

FIG. 33.—The sacrum, from before.
(Drawn by D. Gunn.)

Some of the dorsal vertebræ are peculiar in the arrangement of their facets and demi-facets, while among the *cervical vertebræ* are several whose peculiarities should be more carefully noted. Thus, the first cervical vertebra or *atlas* supports the head and has practically no body, the place of the body being taken by a narrow anterior arch of bone and an opening, continuous with the spinal foramen, into which the *odontoid process of the axis* fits, being held in place by ligaments. At either side on top is a facet for articulation with the occipital bone. There is almost no spine. The second vertebra or *axis* has surmounting the body the odontoid process, with a facet in front for articulation with the atlas and one behind for the transverse ligament to move over. The seventh cervical vertebra or *vertebra prominens* has a very long spinous process—hence name—to which is attached the *ligamentum nuchæ*. It can be felt very distinctly on the living.

Running from the skull down through the spinal column into the sacral vertebræ and formed by the joining of the spinal foramina of the individual vertebræ is an opening called the *spinal canal*, which holds the cord. The cord, however, stops practically at the first lumbar vertebra, where it splits up into the *cauda equina*, only the *filum terminate* extending farther down.

Occasionally the laminæ do not form completely and the membranes of the cord may bulge out and form a tumor, or the cord itself may come out also. This generally occurs in the lumbar region, where it is known as *spina bifida*. If in case of fracture of a vertebra there is paralysis of the parts below due simply to the pressure of a fragment of bone upon the cord, it may be completely cured by removal of the fragment. If, however, the cord suffers injury, the paralysis will remain. *Humpback* or *Pott's disease* is caused by the tubercle bacillus, which eats away the bodies of the vertebræ so that the column caves in and the spinous processes are thrown out in a hump or kyphos.

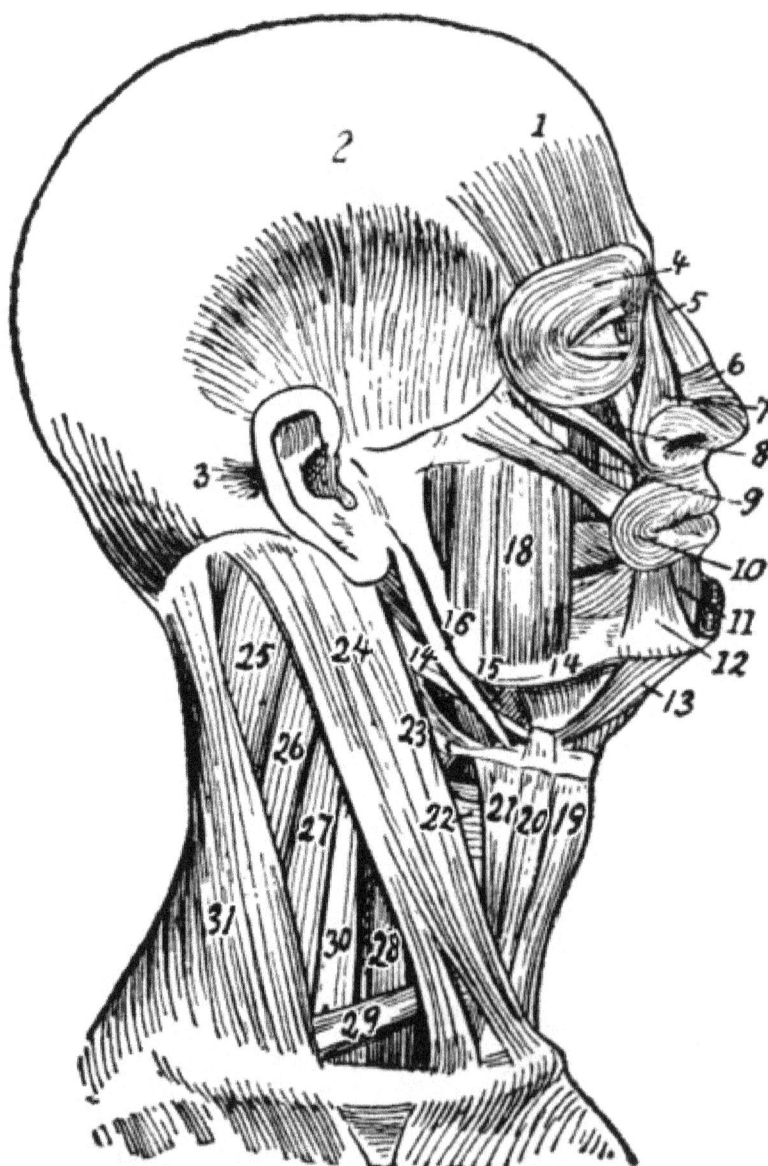

Muscles of the Neck.—Before speaking of the muscles of the back a few of those of the neck had best be taken up. They are numerous but mostly of minor importance. Largest and most important is the *sterno-cleido-mastoid muscle,* which has its origin on the upper part of the sternum and the inner third of the clavicle and is inserted into the mastoid process of the temporal bone. It passes obliquely across the side of the neck and serves to flex the head to the side and to draw the face in the opposite direction. When both muscles contract the head is flexed on the neck and the neck on the chest. In *wry neck* or *torticollis* this muscle is constantly contracted. The *platysma myoides* arises from the fascia over the pectoral, deltoid, and trapezius muscles and is inserted into the lower jaw, the angle of the mouth, and the loose tissue in the lower part of the face. It wrinkles the skin of the neck and depresses the lower jaw. In the cow and horse it is so highly developed that by it the skin can be contracted all over the body to drive off flies. The *rectus capitis anticus major* arises from the third to the sixth cervical vertebræ and is inserted into the occipital bone, serving to flex the head. The *scalenus muscles* have their origin on the lower cervical vertebræ and are inserted into the first and second ribs, thus aiding in the elevation of the ribs as well as in lateral flexion of the neck. The head is held upright by the *ligamentum nuchæ,* which rises from the external occipital protuberance and is inserted into the spinous processes of all the cervical vertebræ except the first.

Muscles of the Back.—The chief back muscles are the trapezius and the latissimus dorsi, which together cover in the back pretty thoroughly. The *trapezius* arises from the occipital bone, the ligamentum nuchæ, and the spinous processes of the seventh cervical and all the dorsal vertebræ and is inserted into the outer third of the clavicle or collar bone and the acromion process and spine of the scapula or shoulder blade. It is thus triangular in shape and covers in the neck and shoulders, serving to draw the head back and to the side. It overlaps the latissimus dorsi.

The _latissimus dorsi_ has its origin by aponeurosis from the spinous processes of the six lower dorsal and all the lumbar and sacral vertebræ, from the crest of the ilium or hip bone, and from the three or four lower ribs, swings across the side, dwindling in size, and is inserted by a small tendon into the bicipital groove of the humerus or upper arm bone, thus covering in the part of the back not covered by the trapezius. It draws the arm down and back, raises the lower ribs, and draws the trunk forward, as in climbing. The flat muscles of the back and abdomen have a tendency to flatten out into aponeuroses, such as occurs in the origin of the latissimus dorsi.

The _levator scapulæ_, from the transverse processes of the upper cervical vertebræ to the posterior border of the scapula, serves to raise the angle of the scapula, and the _rhomboideus major and minor_, from the ligamentum nuchæ, the seventh cervical, and the upper dorsal vertebræ to the root of the spine of the scapula, draw the inferior angle back and up.

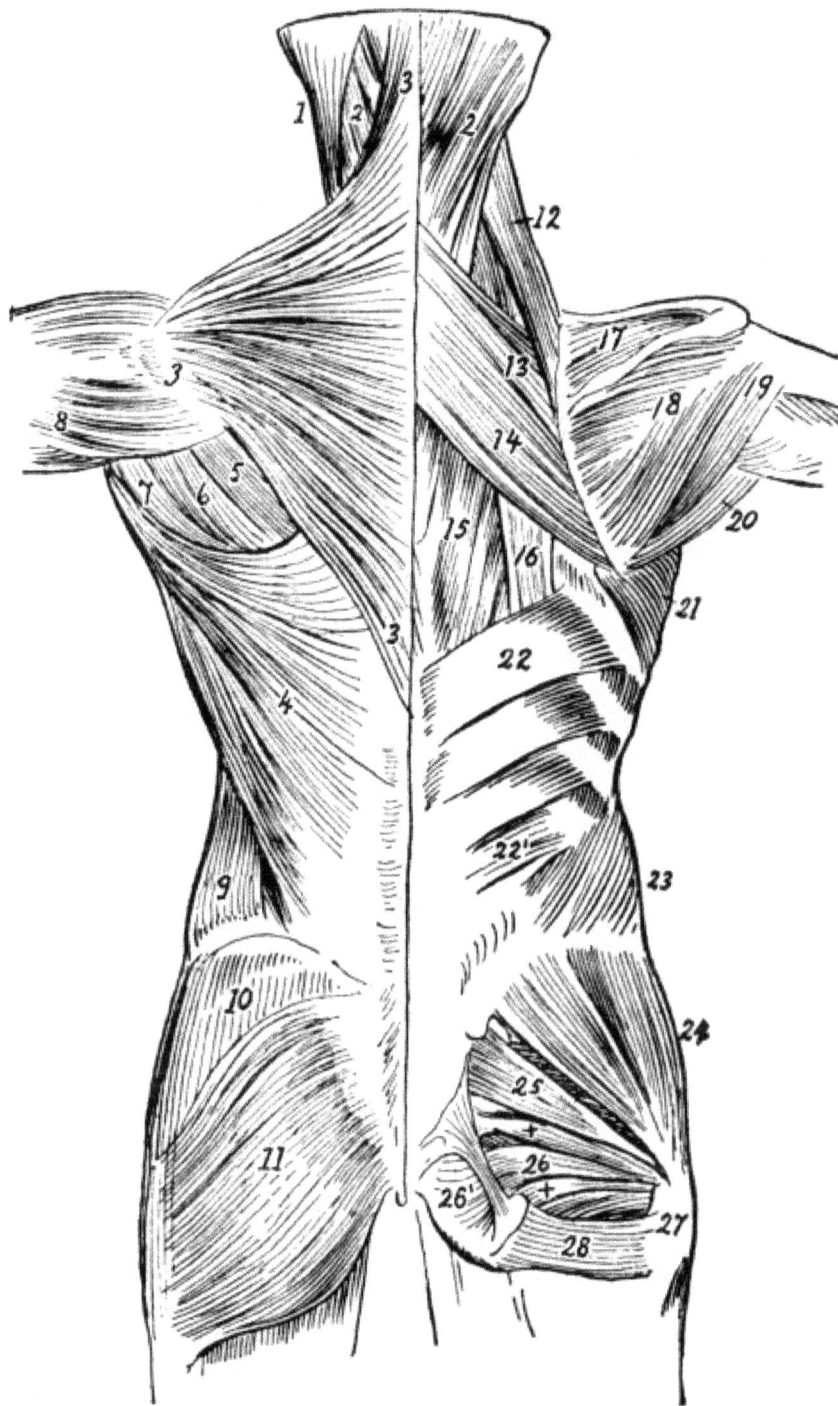

FIG. 35.—Muscles of the trunk from behind (left side, superficial; right side, deep): 1, Sternomastoid; 2, splenius; 3, trapezius; 4, latissimus dorsi; 5, infraspinatus; 6, teres minor; 7, teres major; 8, deltoid; 9, external oblique of abdomen; 10, gluteus medius; 11, gluteus maximus; 12, levator anguliscapulæ; 13, rhomboideus minor; 14, rhomboideus major; 15, part of longissimus dorsi; 16, tendons of insertion of iliocostalis; 17, supraspinatus; 18, infraspinatus; 19, teres minor; 20, teres major; 21, serratus magnus; 22, upper, and 22′, lower

part of serratus posticus inferior; 23, internal oblique; 24, gluteus medius; 25, pyriformis and superior and inferior gemelli; 26, 26´, portions of obturator internus; 27, tendon of obturator internus; 28, quadratus femoris. (Dorland's Dictionary.)

The *blood supply* in the cervical region and about the shoulders comes from branches of the subclavian artery, such as the suprascapular and the transversalis colli. Lower down the supply comes from the posterior branches of the intercostals, dorsal branches of the lumbar, and branches of the internal iliac.

The muscles of the back are supplied by the *spinal nerves*, the spinal accessory also going to the trapezius muscle.

CHAPTER VII.
THE CHEST.

The chest or thorax occupies the upper part of the trunk in front and is a dome-shaped cavity containing and protecting the heart and lungs. Its walls are formed by the dorsal vertebræ at the back, the ribs at either side, and the sternum and costal cartilages in front, all well covered with muscles. The floor is formed by the diaphragm. Through the upper opening of the chest pass the trachea, the esophagus, and many important vessels and nerves.

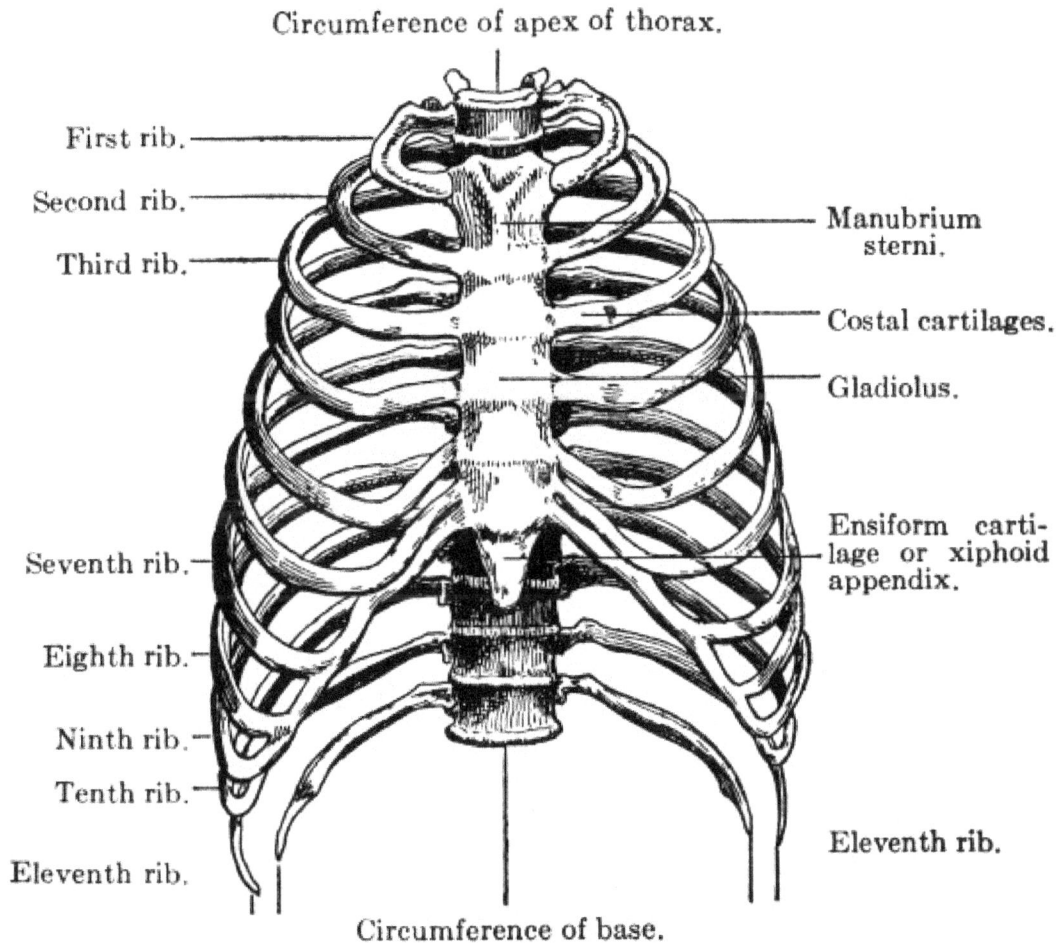

FIG. 36.—Thorax (anterior view.)
(Ingals.)

The shape of the chest may vary in disease. Thus, in rickets there is the prominent "pigeon" breast and the rosary, that is, a bead at the juncture of each rib with the costal cartilage, while in emphysema the chest is enlarged in all directions and barrel-shaped. In severe cases of lateral curvature it is distorted but may be improved by exercises.

The Sternum.—The sternum or breast-bone is a long narrow bone and has three parts, the _manubrium_ or handle above, the _gladiolus_ or sword, and the _ensiform cartilage_ at the lower end. On either side are notches for the costal cartilages; for the first seven ribs as well as the clavicle articulate with it. Except for some muscles along the edges it lies directly under the skin and the ridge between the manubrium and the gladiolus can be felt in the living, a fact which assists in determining the position of the different ribs in cases of fracture, as the second rib articulates at this point.

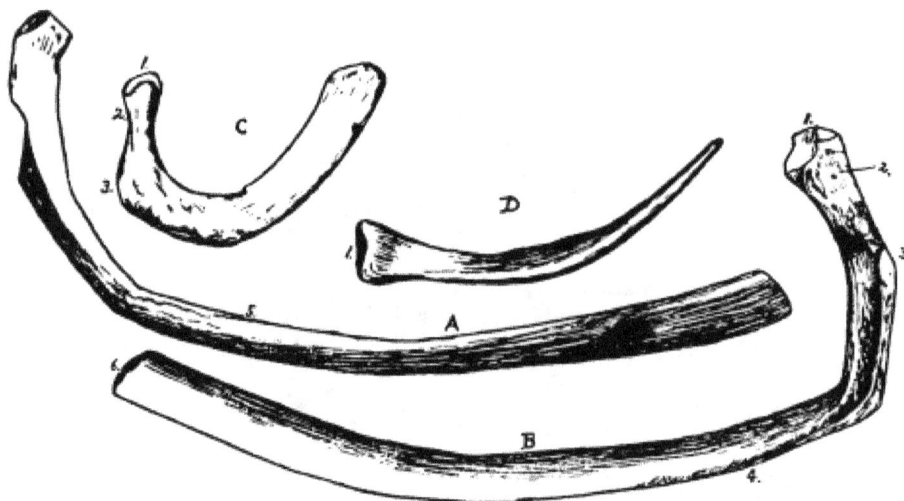

FIG. 37.—A and B, typical ribs; C, first rib; D, twelfth rib. 1, head; 2, neck; 3, tuberosity; 4, grooved edge; 5, shaft; 6, oval depression for costa cartilage.

The Ribs.—The ribs are twenty-four in number, twelve on each side, of which the upper seven, which articulate with the sternum by individual cartilages, are called _true ribs_, the other five _false ribs_. Of the false ribs the upper three articulate indirectly with the sternum through the seventh cartilage, with which their cartilages unite, while the other two have their anterior extremities free and are called _floating ribs_. All the ribs slope down toward the front and are by nature more freely movable in women than in men. Most of the ribs have a head divided by a little ridge into two facets for articulation with the dorsal vertebræ, a flattened neck, a tuberosity at the base of the neck with a facet for articulation with the transverse process of the vertebra below, an angle, and a shaft, which is externally convex and is grooved on its lower edge for the intercostal vessels and nerve. The first and second, eleventh and twelfth ribs, however, are somewhat peculiar, the first two being shorter, flatter and rather broader than the rest and the first having only one facet on the head, while the last two have only one facet on the head and no neck or tuberosity.

The _costal cartilages_ serve to prolong the ribs and greatly increase the elasticity of the chest wall. They grow longer down to the seventh and then decrease again in length.

The ribs, except the first and second, which are protected by the clavicle, are frequently _broken_. Such a break causes pain in breathing and sometimes the end of a rib pierces the lung tissue and swelling all over the body results, due to the presence of air. _Caries_ or death of the rib is also frequent. _Fracture of the sternum_ occurs occasionally, generally from direct force, as from a blow with the knee in foot-ball, and there may be dislocation between the manubrium and gladiolus.

Muscles of the Chest.—The spaces between the ribs, from the tubercle of the rib behind to the cartilage in front, are filled by the _external intercostal muscles_, which pass downward and forward from the lower border of one rib to the upper border of the one below. There are, therefore, eleven pairs of these muscles. There are also eleven pairs of the _internal intercostals_, which commence at the sternum and extend back to the angle of the rib. These extend downward and backward. The external intercostals raise and evert the ribs in inspiration, the internal depress and invert them in expiration.

The chief respiratory muscle, however, is the _diaphragm_, a somewhat fan-shaped muscle that forms the floor of the chest cavity. It takes its origin from the ensiform cartilage, the six or seven lower ribs and their cartilages, and from the upper three or four lumbar vertebræ, that is, from the whole of the internal circumference of the thorax, and is inserted into the central cordiform tendon. It has several large and several small openings for the aorta, the esophagus, the venæ cavæ, the thoracic duct, and various nerves, and its surfaces are covered by serous membranes, by the two pleuræ and the pericardium above and by the peritoneum below. It partially supports the heart and lungs. Convex toward the chest, it becomes flattened in contraction and so increases the capacity of the chest. It aids in all expulsive acts, as sneezing, coughing, laughing, urinating, defecating, vomiting, and childbirth. *Hiccough* is spasm of the diaphragm.

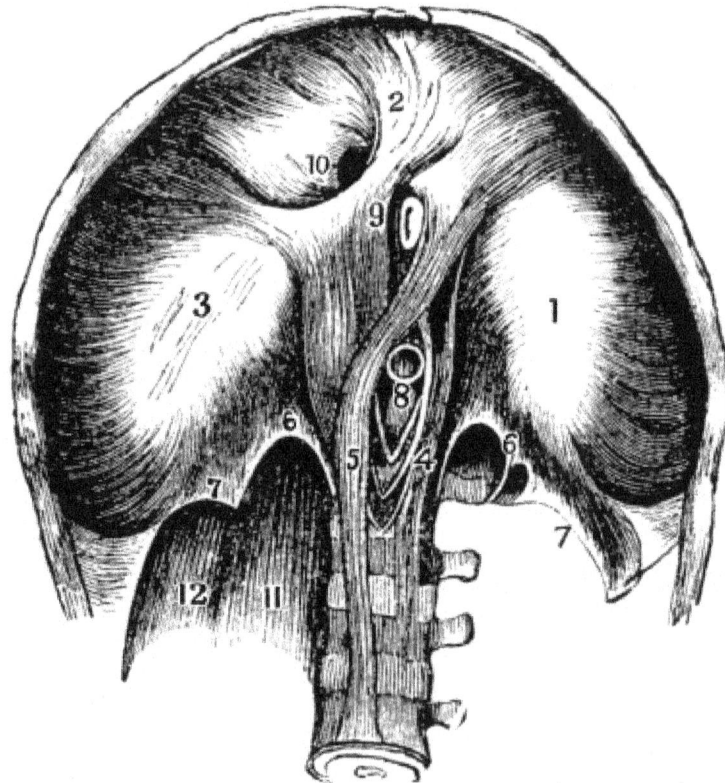

FIG. 38.—Interior view of the diaphragm. (Leidy.) 1-3, The three lobes of the central tendon, surrounded by the fleshy fasciculi derived from the inferior margin of the thorax; 4, 5, the crura; 6, 7, the arcuate ligaments; 8, aortic orifice; 9, esophageal orifice; 10, quadrate foramen; 11, psoas muscle; 12, quadrate lumbar muscle.

The *arteries* of the chest are the intercostal branches of the subclavian and the thoracic aorta, the phrenic, mediastinal, and intercostal branches of the internal mammary, and the thoracic branches of the axillary.

The *nerves* are the intercostals and phrenics.

Mammary Glands.—On the outside of the chest walls, lodged in the fascia of the pectoral muscles, are the mammary glands, accessory organs of the generative system. They exist in both sexes but are only rudimentary in the male. In the female they are small before puberty but enlarge as the generative organs become more completely developed, forming two hemispherical eminences, one on either side, between the third and seventh ribs. During pregnancy they increase once more in size preparatory to the secretion of the milk, and in old age they atrophy. From the middle projects a small pinkish-brown conical eminence, the *nipple*, surrounded by a paler area, the *areola*. After the second month of pregnancy both nipple and areola become darker in color, a point of great diagnostic value in early pregnancy.

The mammary glands themselves consist of lobules of gland tissue with a central lactiferous tubule, the lobules being gathered into lobes with fatty tissue between. From the juncture of these tubules result fifteen or twenty excretory ducts, the *tubuli lactiferi*, which converge toward the areola. Beneath the nipple they dilate, forming the *ampullæ*, and then contract again to pass out through the nipple as straight tubes.

Breast abscess occurs most commonly in nursing mothers, as where a part is most active there is most danger of abscess. Many benign *tumors* of the breast, as the fibrous tumors, occur and are especially common in young women. If a fibrous tumor is allowed to develop it may become cancerous. *Cancer*, however, generally occurs after the age of forty and is usually due to some irritation, as to a blow from a ball.

The *arteries* of the breasts are the thoracic branches of the axillary, the intercostal, and the internal mammary.

The *nerves* are from the thoracic cutaneous.

CHAPTER VIII.
THE HEART AND CIRCULATION.

The Heart.—Shielded within the chest are, as has been said, the heart and lungs. The heart lies on the left side behind the sternum and the cartilages of the fourth to seventh ribs in a closed, conical, membranous sac, the *pericardium*, which is attached by its base to the central tendon of the diaphragm, and whose point extends up between the pleuræ of the lungs. This sac has an external fibrous layer and an internal serous layer that is reflected back over the heart itself, forming a closed sac, within which a thin fluid is secreted that serves to reduce friction during the movements of the heart, the two inner surfaces sliding over each other with every beat.

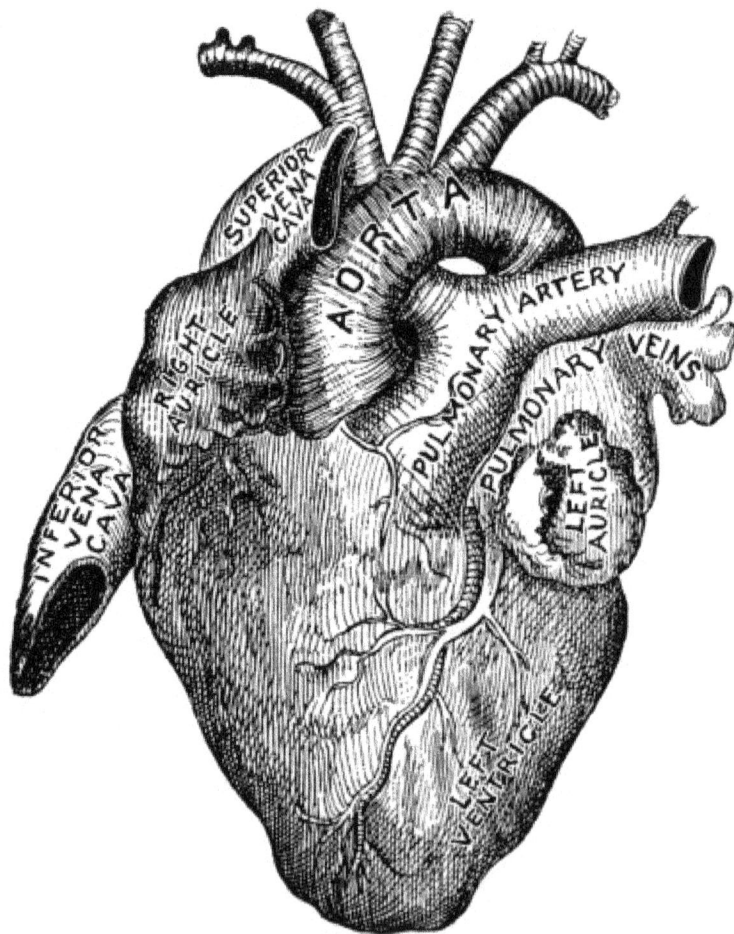

FIG. 39.—The heart.
(Stoney.)

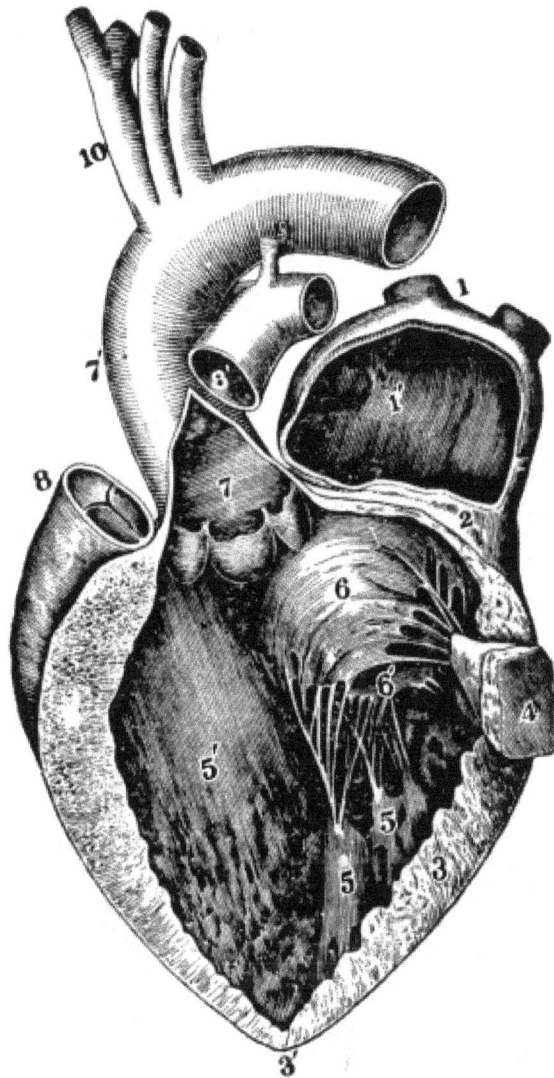

FIG. 40.—Left auricle and ventricle, opened and part of their walls removed to show their cavities: 1, Right pulmonary vein cut short; 1´, cavity of left auricle; 3, 3´, thick wall of left ventricle; 4, portion of same with papillary muscle attached; 5, the other papillary muscles; 6, 6´, the segments of the mitral valve; 7, in aorta is placed over the semilunar valves; 8, pulmonary artery; 10, aorta and its branches. (Allen Thomson.)

The heart itself is a hollow conical organ composed of cardiac muscle, a combination of smooth and striated fibers found nowhere else in the body. It lies obliquely, base up, between the lungs, suspended by the great blood-vessels and with the apex directed downward, forward, and to the left, the apex beat being normally felt in the fifth intercostal space, one inch inside and two inches below the left nipple. In size it varies in different people and is generally smaller in women than in men. On the average it is five inches long, three and a half inches broad, and two inches thick. A man's heart usually weighs about eleven ounces and that of a woman nine ounces. It never leaks except from disease and such leakage is fatal.

The Cavities.—The heart contains four cavities, two <u>auricles</u> above and two <u>ventricles</u> below, with a longitudinal septum between the auricle and ventricle on the right and those on the left. The posterior surface is largely made up of the left ventricle and the anterior of the right ventricle. The right auricle, which receives the blood from the general circulation, has a capacity of about two fluid ounces and is larger than the left, which receives the blood returning from the lungs, though its walls are thinner. Of the ventricles the left is the larger and its walls are about three times as thick as those of the right, for it has to send the blood all over the body. All the cavities are lined with smooth, transparent, serous membrane, the *endocardium*, which is continuous with the intima of the great vessels.

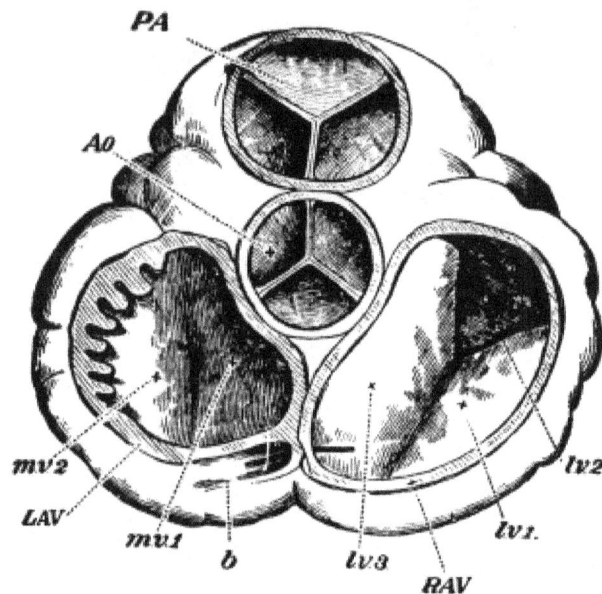

FIG. 41.—Orifices of the heart, seen from above, both the auricles and the great vessels being removed: *PA*, Pulmonary artery and its semilunar valves; *Ao*, aorta and its valves; *RAV*, tricuspid, and *LAV*, bicuspid valves; *mv*, segments of mitral valve; *lv*, segment of tricuspid valve. (Huxley.)

The Valves.—The opening from the auricle into the ventricle on either side is guarded on the ventral side by a valve formed of folds of endocardium. The valve on the right side has three flaps or cusps and is called the _tricuspid valve_, while that on the left has two flaps, larger and thicker than those of the tricuspid, and is known as the _bicuspid or mitral valve_. The flaps of either valve are kept from being forced into the auricle in closing by fine tendinous cords, the _chordæ tendineæ_, which are attached to the _columnæ carneæ_, muscular bands or columns projecting from the walls of the ventricle, which contract and hold the chordæ tendineæ taut. The opening into the pulmonary artery is from the posterior part of the right ventricle and is guarded by the _semilunar or pulmonary valve_, while the aortic opening from the left ventricle is guarded by a similar valve, the _aortic valve_, the most important valve in the body. All these valves are planned primarily to prevent regurgitation of the blood during contraction of the heart muscle. Pressure in the ventricle must exceed that in the arteries before the semilunar valves will open and the blood can be driven out, just as the auriculo-ventricular valves remain closed until the pressure in the auricles exceeds that in the ventricles.

The _heart beat_ is caused by the twisting of the heart upon its axis during contraction of the muscle. Normally it beats rhythmically and regularly, whatever a person does, at a rate of about seventy-two contractions to the minute in the adult. To the regular _cardiac cycle_, as it is called, there are two periods, the _systole_ and the _diastole_, the former representing the period of contraction of the ventricles, when the blood is sent to the lungs and over the body, and the latter representing the period of rest following the emptying of the ventricles, during which they are refilled. Contraction of the heart occupies one-fifth of the time of one beat, dilatation two-fifths, and the pause two-fifths. There are really two systoles, one of the auricles and one of the ventricles, but they come so close together that they are practically simultaneous so far as sound is concerned, though they can be distinguished by sight. During systole the tricuspid and mitral valves close sharply to prevent regurgitation into the auricles, while the semilunar valves open to let the blood out. The cardiac cycle is, therefore, as follows:

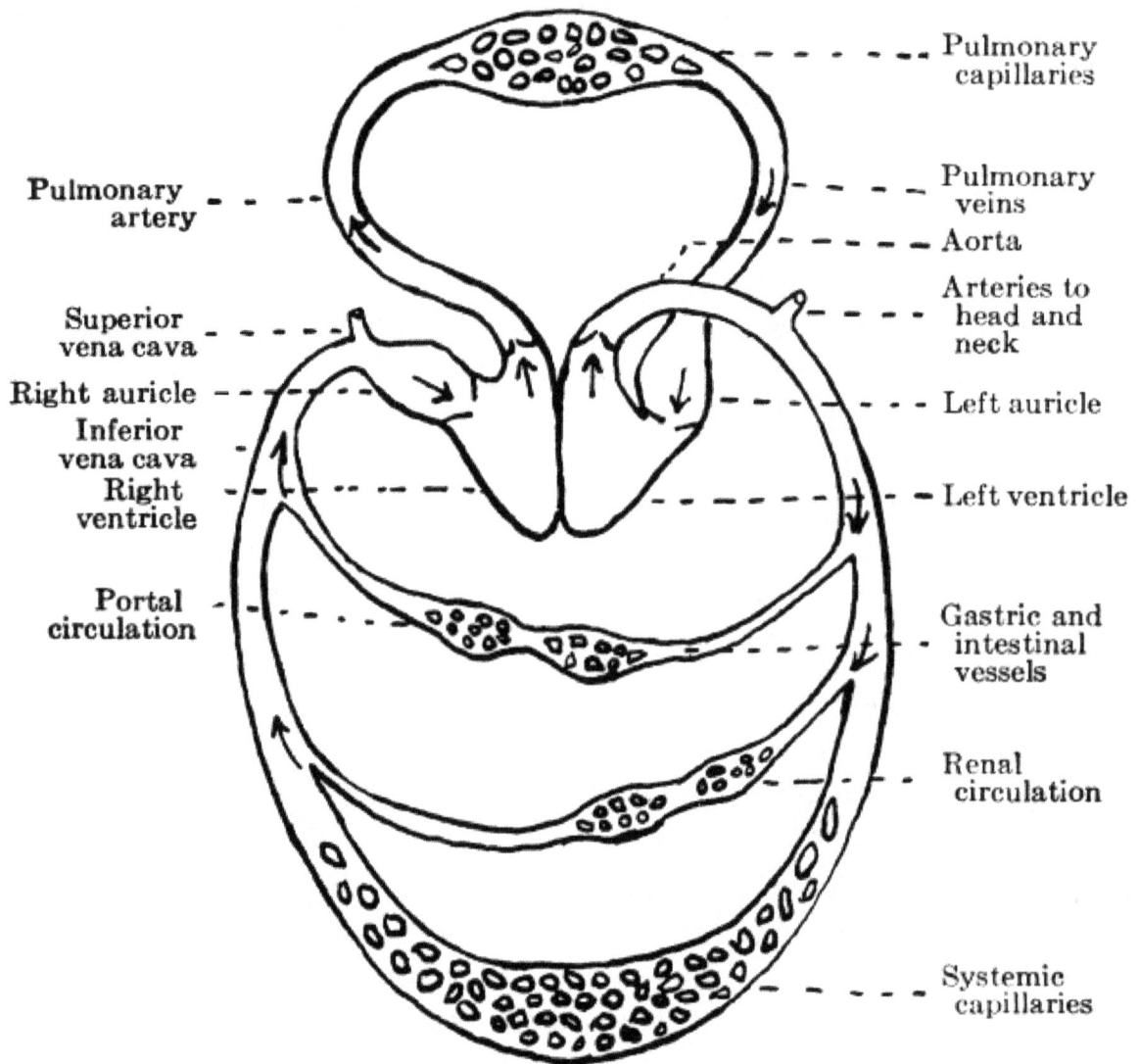

FIG. 42.—Diagram of the circulation.
(After Kirke.)

Circulation.—The blood, after it has given off its oxygen and collected carbon dioxide, returns to the heart through two main channels, the *superior* and *inferior venæ cavæ*, the former bringing the blood from the upper part of the body, including the head, neck, and arms, and the latter from the lower part below the diaphragm. The two vessels empty along with the *coronary sinus*, which is guarded by the *coronary valve*, into the right auricle. At the same time that they empty into this auricle the four *pulmonary veins*, the only veins that carry arterial or oxygenated blood, are emptying the fresh blood from the lungs into the left auricle. When both auricles are full, they contract and send the blood into the ventricles, the *auricular systole*. As the blood comes through into the ventricles it probably comes around by the walls and closes the auriculo-ventricular valves, though just how the valves close is not certain. When the two ventricles are full they in turn contract, the *ventricular systole*, and the blood is forced out, that in the right ventricle passing to the lungs for its new supply of oxygen through the *pulmonary artery*, the only artery to carry venous blood, and that from the left ventricle entering the aorta for general distribution through the body. Following the systole is a pause, the *diastole*, while the heart fills again.

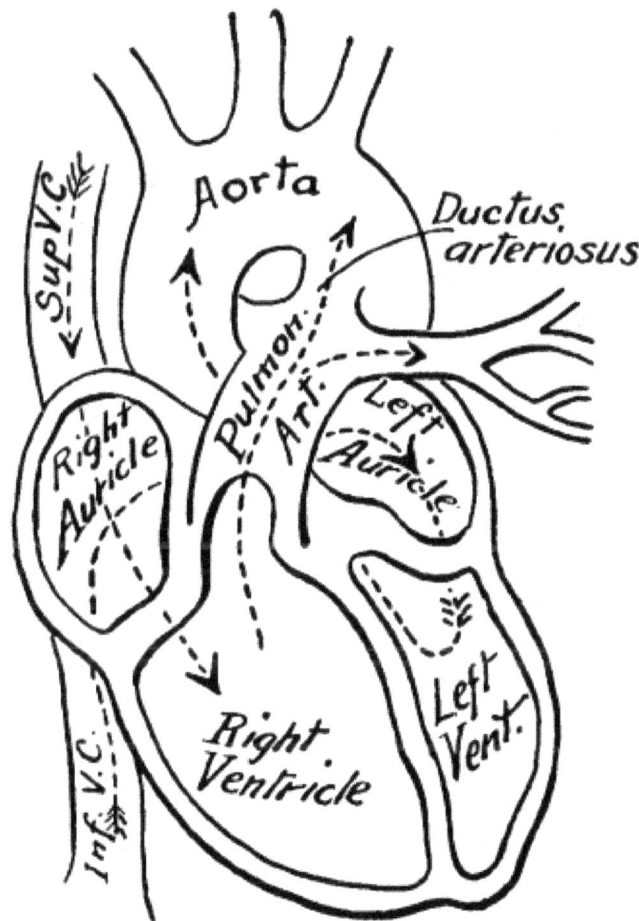

Fig. 43.—The fetal circulation.

Circulation in Fetus.—In the fetus there is direct communication between the two auricles through the *foramen ovale*, which normally closes at birth, though occasionally it remains open. There is also communication between the pulmonary artery and the arch of the aorta through the *ductus arteriosus*. The freshly oxidized blood comes to the fetus through the placenta, from which it is brought along the umbilical cord in the umbilical vein to the liver and thence to the inferior vena cava, where it mixes with the blood from the lower extremities. By the inferior vena cava it is carried to the right auricle, where the *Eustachian valve*—a valve between the inferior vena cava and the auriculo-ventricular opening, larger in the fetus than in later life where it serves no special purpose—guides it across the auricle and through the foramen ovale to the left auricle. From this auricle, together with a small amount of blood from the lungs, it goes to the left ventricle and is distributed by the aorta almost entirely to the head and upper extremities. Hence their large size and perfect development at birth. Returned from the upper extremities by the superior vena cava, the blood enters the right auricle again and, passing over the Eustachian valve this time, descends to the right ventricle, from which the greater part passes by the pulmonary artery and the ductus arteriosus to the descending aorta, though a small amount keeps on through the pulmonary artery to the lungs. In the aorta it mixes with the blood from the left ventricle and part goes to supply the lower extremities, though the greater part is carried back to the placenta through the two umbilical arteries. The fact that the greater part of the blood traverses the liver accounts for its large size at birth, while the lower extremities, which receive for the most part blood that has already circulated through the upper extremities, are of small size and imperfectly developed.

Arteries.—After birth the arterial blood for the general circulation leaves the heart by the aorta, the main distributing artery of the body. Through this and its branches it is carried throughout the body in what, with the return of the venous blood by the venæ cavæ and other smaller veins, is known as the *systemic circulation*. The *aorta* ascends from the left ventricle and arches backward to the left over the root of the left lung to descend along the spinal column at the left to the fourth lumbar vertebra, about opposite the umbilicus, where, considerably diminished in size by the branches it has given off, it divides into the two *common iliacs*. For convenience its different parts are named, according to their position, the *ascending aorta*, the *arch of the aorta*, and the *descending aorta*, the last being subdivided into the *thoracic* and the *abdominal aorta*.

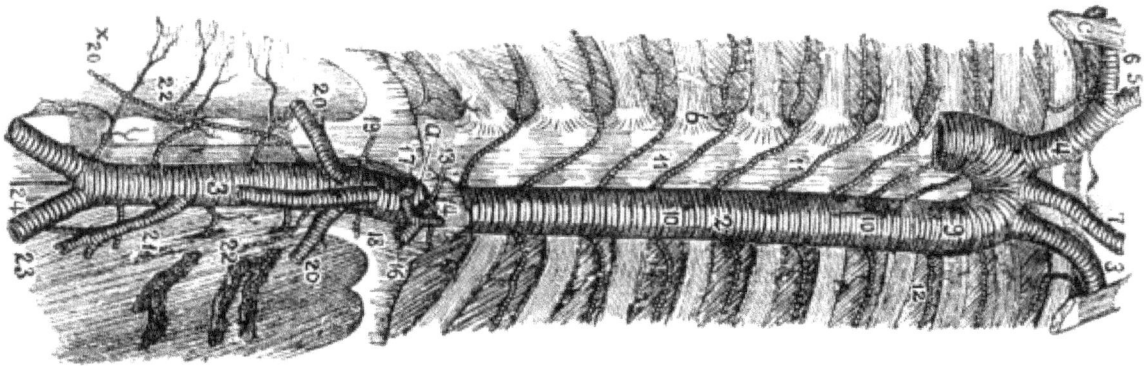

FIG. 44.—The aortæ and their branches.
(Leidy.)

From the ascending aorta come off the *coronary arteries* which supply the heart muscle itself, as the coronary sinuses carry off the venous blood from the heart. From the arch are given off the *left common carotid* and *left subclavian* and the *innominate*, which divides into the *right common carotid* and *right subclavian*.

The *common carotids* pass up the neck behind the sterno-cleido-mastoid muscles in a line from the sterno-clavicular joint to a point midway between the mastoid process and the angle of the lower jaw and divide opposite the upper border of the thyroid cartilage into the *internal* and *external carotids*, of which the former with its branches supplies the anterior part of the brain, the eye and forehead, and the latter the neck and face.

The *subclavian* is the artery of the upper extremity but its vertebral branch goes to the brain, where with its fellow it forms the *basilar artery*, whose branches together with the branches of the internal carotid form the *circle of Willis* at the base of the brain. Other branches of the subclavian are the *thyroid axis*, with branches to the neck and shoulders; the *internal mammary*, with branches to the chest walls, mediastinum, and diaphragm, such as the musculo-phrenic and superior epigastric; and the *superior intercostal*. At the lower border of the first rib, over which it passes, the name *axillary* is substituted for *subclavian*, while at the lower border of the axilla, where it starts down the arm, it is called the *brachial artery*. At the elbow the brachial divides into the *radial* and *ulnar arteries*. The *axillary artery* sends branches to the chest and shoulder and is more frequently injured than any other artery except the popliteal. *Aneurism* may occur in it and is very likely to occur in the thoracic aorta.

From the *thoracic aorta* branches go to various of the chest contents, while the *abdominal aorta* supplies the abdominal viscera. Among the branches of the abdominal aorta are: the *celiac axis*, which has a gastric, an hepatic, and a splenic branch; the *superior* and *inferior mesenteric* to the intestines; the *renal*; the *suprarenal*; the *spermatic* or *ovarian*; the *inferior phrenic*; and the *lumbar*.

The *common iliacs* divide at the upper edge of the sacrum into the *external* and *internal iliacs*, of which the latter with its branches supplies the walls and viscera of the pelvis and the inner part of the thigh. The external iliac and its branches go to the thigh, leg, and foot.

Veins.—Of the veins few need be mentioned by name. The deep veins have the same names as the arteries they accompany, though there are two *innominate veins* where there is only one innominate artery, the *subclavian* and *internal jugular* veins on either side joining to form an innominate vein and the two innominates in turn forming the *superior vena cava*. Of the superficial veins the *external* and *internal jugular* correspond to the common carotid arteries and return the blood from the head and face. The external jugular vein is important because it is the largest superficial vein in the neck and is often cut in suicide. The *median vein* is found at the bend of the elbow and is used in letting blood and in giving salt solution, while the *basilic* is on the inner side and the *median cephalic* on the outer side of the upper arm. *Varicosity* often occurs in the *internal* or *long saphenous* and the *external* or *short saphenous* in the leg. The *inferior vena cava* is formed by the juncture of the two *common iliac veins*.

Portal Circulation.—The portal system of veins includes four large trunks which collect the blood from the viscera of digestion, the *superior* and *inferior mesenteric veins* from the intestines, the *splenic vein* from the spleen, and the *gastric* from the stomach. These join together to form the *portal vein*, the only vein that breaks up into capillaries. This divides and ramifies through the liver, whence it emerges as the *hepatic veins*. The whole is known as the portal circulation.

Pulmonary Circulation.—Of the pulmonary circulation and its vessels a few words might also be said. The *pulmonary artery*, which carries the blood from the right ventricle to the lungs, is only about two inches long and divides into a right and a left pulmonary artery, which pierce the pericardium and go to their respective lungs. The right one is the larger and longer, for it has farther to go and gives off a branch to supply the third lobe of the right lung. The vessels finally divide and subdivide, terminating in the *pulmonary capillaries*. The venous capillaries then gather together to form a main vein in each lobule, these veins uniting into two trunks for each lung, the *pulmonary veins*, which empty into the left auricle.

Nerves of Heart.—The muscular fibers of the heart have the power of rhythmical contraction. Independent nerve centers or ganglia are also found in the muscular walls and influence the mechanism of the heart, especially the acceleratory mechanism. Thus, in some of the lower animals the heart can be removed from the body, and if placed in normal salt solution will go on beating for some time. The heart is controlled, however, by two nerves, the *vagus* or *pneumogastric* and the *sympathetic*. Of these the vagus is the inhibitory mechanism. It acts as a check and makes the heart's action regular and rhythmic. If it is cut, the action of the heart becomes very rapid and irregular. The sympathetic is the acceleratory mechanism. When the vagus alone is stimulated, it first slows, then stops the heart, for it weakens the systole and prolongs diastole. Acceleration follows stimulation of the sympathetic, both the rapidity and the force of the beat being increased. When a person faints from a blow in the abdomen, it is because the pneumogastric is affected and inhibits the action of the heart. The work of the heart is very dependent upon its nervous condition and functional diseases of the heart are practically wholly due to nervous derangement.

Heart Sounds.—Through the stethoscope two heart sounds may be heard. They are known as the first and second sounds. The first is a soft, rushing sound, stronger and louder than the other, and is caused in part by the contraction of the muscle itself when the blood is forced out and in part by the closure of the auriculo-ventricular valves. The second sound is shorter and sharper, a snap, and is caused by the closure of the semilunar valves when the contraction of the ventricles ceases and they begin to refill. In certain diseased conditions, where the edges of the valves are roughened, they do not snap properly and the sound varies from the normal.

The Heart Beat.—The rate of the heart beat is proportionate to the size of the person and increases in rapidity as the size diminishes. If the ear is placed over the abdomen of a pregnant woman, the heart of the fetus can be heard beating very rapidly. In prolonged labor it may become more rapid or very faint and warn the doctor that something should be done. The usual rate of the pulse in the fetus is 140 to 150 times a minute, though it varies with size and sex. At birth it drops to 140 to 130; for the first year it is 130 to 115; for the second year 115 to 105; for the third year 105 to 95; from the seventh to the fourteenth years 80 to 90; from the fourteenth to the twenty-first years 75 to 80; from twenty-one to sixty 60 to 75. In old age it rises a little and is 75 to 80. The rate is higher in the average woman than in the average man and increases with exercise, with increase of temperature, and in high altitudes, where the atmospheric pressure is less.

At each beat of the heart from four to six ounces of blood are expelled into the pulmonary artery and the aorta, and in 22 or 23 beats all the blood in the body passes through the heart. The power exerted by the heart every minute in thus driving the blood upon its course has been estimated as sufficient to raise its own weight, three-quarters of a pound, the height of the Washington monument or 150 meters; for the ventricles have to force the blood into vessels already full.

Factors Affecting Circulation.—There are three main factors in the circulation: 1. the systole, which gives the blood its first impulse; 2. the peripheral resistance in the capillaries, which serves to hold it in check, slowing the circulation and doing away with its rhythmic character, and 3. the elasticity of the walls of the arteries.

If a ligature is tied about an artery, there is a swelling on the side toward the heart, while in the case of a vein, the swelling is on the side away from the heart, that is, the swelling is in either case on the side from which the blood comes. When an artery is cut, however, the blood comes out rhythmically in spurts, though from a cut vein it oozes slowly and regularly. For the blood is pumped out by the heart rhythmically and its rhythmic beating against the walls of the artery is felt in the *pulse*, which follows slightly after the beat of the heart itself. The pulse is due to the fact that the vessels into which the blood is forced are already full. This causes a local dilation at the beginning of the artery which passes with diminishing force along its entire length, the distention being due to the fact that more force is needed to drive the blood through the small arteries and capillaries than to stretch the elastic walls of the aorta and the large arteries. It is this elastic character of the arteries that makes the blood flow constant, for otherwise the blood would come intermittently in jets, as it is pumped from the heart. The elastic walls of the vessels, however, offer a certain resistance to the pumping of the fluid through them and at the same time, by relaxing between whiles, allow a certain amount of fluid to be retained in them, so that they continue full and the flow is more or less constant. The insufficient outlet also helps to make the flow constant.

By the time the blood reaches the veins its rhythmic character has been done away with, but though there are no elastic walls in the veins, it still has force enough after the slowing in the capillaries to return to the heart. In this it is aided to a certain extent by the valves and by the action of the skeletal muscles as they contract and expand, especially in the arms and legs, where the blood runs perpendicularly and there is a high column to be supported. There are also more veins than arteries, each large artery having two large veins, the venæ comites, to help get the blood back to the heart, and the veins anastomose freely. Thus, if the blood cannot get back by one channel it does by another. In parts like the brain, where it is very important that there should be no compression, since any disturbance of circulation would lead to serious results, the vessels are enclosed in thick walls, and in the liver, through which all the blood passes and where compression is sure to cause trouble, the veins are simply caverns carved out in the organ and have no walls. They lie open when the organ is opened. *Varicose veins* are the result of valves giving way through inherited weakness or disease so that others have an unduly large weight to support.

The Pulse.—The *pulse wave* is characterized by a quick rise and a slow fall, though this cannot ordinarily be distinguished by the finger. In some slow fevers, however, the fall is very long and distinct ripples can be felt. This is known as the *dicrotic pulse*. With age the arterial walls grow stiffer and more rigid and less adapted to their work. In certain cases of heart disease the heart does not transmit all the beats to the pulse and to get the true rate the heart must be listened to.

The rate at which the pulse wave travels varies with the size of the artery and the force of the heart beat but is about 15 to 20 feet a second. The flow is most rapid in the arteries because they are nearest the heart, where the pressure is greatest, and slowest in the capillaries, where the area is greatest, the sectional area of the capillaries, known as the *peripheral area* because it is farthest from the heart, being larger than that of the large arteries. Thus rapidity of flow varies with pressure and with area.

Blood Pressure.—Liquids, moreover, are incompressible and exert pressure on the walls of the tubes through which they pass. The amount of pressure depends upon the inflow and outflow, increasing directly with the inflow and inversely with the outflow, that is, the smaller the outlet the greater the pressure, and *vice versa*. The pressure is also greatest nearest to the inflow and gradually decreases with distance until at the point of outflow there is practically no pressure. So, in the arteries the blood pressure is greatest in the large vessels nearer the heart and gradually decreases as they branch into smaller and smaller vessels. In passing through the capillaries, owing to their small size and resultant increased friction, the blood meets with more resistance, the peripheral resistance, and this resistance usually regulates the pressure in the arteries. The greater the peripheral resistance, as a rule, the greater the arterial pressure. The pressure in the capillaries is very slight and in the veins there is practically no pressure. In fact, in the large veins near the heart the pressure is negative and the blood is almost sucked into the heart.

Pressure, then, is greatest in the arteries and least in the veins, while the rate of flow is fastest in the arteries—300 to 500 millimeters a second—and slowest in the capillaries—75 millimeters a second—being a little faster again in the veins—200 millimeters a second.

Blood pressure is gauged by opening a vessel and inserting a *manometer,* the pressure being determined by the height to which the mercury is raised. In man the pressure in the arteries is 120 to 160 millimeters. It is considerably heightened during inspiration by the increased pressure of the lungs on the heart and great vessels. In pericarditis the opposite is true.

When the blood pressure is high, the *pulse* is small and travels fast, because the wall of the artery is already highly stretched. Such a pulse is hard and incompressible. A large pulse occurs where the heart is strong and the pressure is low, owing to peripheral dilatation. A low-pressure pulse is soft and compressible if the heart beat is weak. A slow pulse is generally stronger than a rapid one.

The *nerve supply of the blood-vessels* comes from the spinal cord through the vasomotor nerves, which are connected with the sympathetic system and are distributed to the smooth muscle fibers of the vessels. They are of two classes, the *vasoconstrictors*, which diminish the lumen of the vessels, and the *vasodilators*, which increase the size of the vessels. By these nerves the general tone of the arteries is kept up. They are distributed chiefly to vessels in the skin and in the abdominal organs and the constrictors are probably the more important. When the constrictors are stimulated, three phenomena occur: 1. diminished flow through the vessel, due to its diminished size; 2. increased general arterial pressure, and 3. increased flow through the other arteries. When the dilators are stimulated the opposite effect is produced: 1. the flow through the vessel is increased; 2. there is decreased arterial pressure, and 3. there is decreased flow through the other arteries. The palor of fright is due to the action of the vasoconstrictor nerves of the face and blushing to the action of the vasodilators. Heat stimulates the vasodilators so that more blood goes to the skin, perspiration begins, and the body is cooled by evaporation. Cold stimulates the vasoconstrictors and the blood is kept within the body, where it cannot cool. If a part has too much blood, an impulse passes by the vasoconstrictors to lessen the supply, while if more blood is needed a message goes to the central nervous system and an impulse passes by the vasodilators to flush the organ. The more active a part is in functioning the greater the number of capillaries, except in the brain, which has only large vessels. The vessels of the intestines contain much blood and are capable of containing all the blood in the body.

The Blood.—The blood itself, which thus circulates through the body, carrying nutrition to the tissues and removing waste, is a complex fluid of a bright red color. Its amount has been calculated to be about one-thirteenth of the body weight. One-fourth of it is generally in the heart, lungs, and large arteries and veins, one-fourth in the liver, one-fourth in the skeletal muscles, and one-fourth variously distributed through the other organs. If there is too little blood, the vital processes cannot go on as they should, while too great a supply causes weakness rather than strength. So the tendency is to keep the amount constant and any blood added is disposed of and any blood lost is replaced. In starvation it is the last tissue to be used up, for on it the life of the other tissues depends.

Composition.—In composition the blood is practically the same in all arteries and fundamentally the same everywhere, but in passing through certain organs certain substances are added to or taken from it, so that its character changes more or less. Thus it varies somewhat in composition in different parts of the body, as in the liver and kidneys. It has five main functions: 1. the conveying of fuel from the digestive tract to the tissues, or force production; 2. the carrying of oxygen to the tissues; 3. the carrying of tissue-building materials, or tissue building; 4. the distribution of heat; and 5. the removal of waste products.

The blood is slightly alkaline in reaction, of a saltish taste, and has a specific gravity of 1055. Its temperature is about 100° Fahrenheit or 37.8° Centigrade. It is made up of two parts, the *plasma* or fluid portion and the <u>corpuscles</u> or solid portion. The plasma, again, which is transparent and almost colorless, consists of two materials, the *blood serum* and *fibrin*. Fibrin does not exist as such in the body nor in freshly shed blood, but there is a substance named *fibrinogen* which is worked on by another substance, the *fibrin ferment*, to form fibrin. Both fibrin ferment and fibrinogen can be isolated from the blood.

Coagulability.—In the body the blood is perfectly fluid and under normal conditions does not coagulate. But, though fluid when first shed, upon standing it gradually becomes viscid, that is, in two or three minutes, then jelly-like, in five to ten minutes, and grows firmer and firmer until there finally appears around this jelly-like mass or clot a yellowish fluid, the *serum*. The *clot* is made up of the corpuscles and fibrin. If some blood is drawn and set on ice until the corpuscles settle, the plasma can then be drawn off, and after it has stood a while in a warm place coagulation will take place, a mass of fibrin forming in the middle. It takes from one to two hours for clotting to be complete. In very slow clotting at a low temperature the white corpuscles appear in a layer on top of the clot, the buffy coat.

Of *fibrin* little is known, but its formation is the most important step in clotting, as its presence is absolutely essential. If it is removed by whipping, the blood will not clot. It is a delicate, stringy material, elastic and contractile, and contains certain salts of lime and magnesium, upon whose presence its power of coagulation depends. The coagulability of blood differs in different people and is occasionally so little as to make operation dangerous.

The most favorable temperature for clotting is that of the body, extreme heat preventing it and cold delaying it. That the blood does not clot in the body must be due to some relation between the blood and the walls of the arteries and veins that prevents it, just as the walls of the stomach are not digested by the juices secreted. Though coagulation does not normally take place in the body, it does take place when a blood-vessel is injured or when the blood comes in contact with the air, a wise provision of nature, as otherwise the tendency would be for bleeding to go on indefinitely after injury. The greater the surface with which the blood comes in contact the more quickly it clots. Injury to the vessel wall itself is necessary; the endothelium must be cracked. Under extreme injury the muscular coat of the vessel undergoes spasmodic contraction and partially closes it. Hence a wound caused by tearing is less likely to bleed than one due to cutting.

The valves of the heart, which are covered with endothelium, are frequently the seat of fibrin coagulation, bits of the fibrin thus formed giving rise to conditions in various kinds of heart trouble. Or the bits of fibrin float in the blood and perhaps lodge in the small vessels of the brain and cause apoplexy. Pus in various parts of the body will set up coagulation in nearby arteries. In fact, the presence of any foreign substance in the blood causes clotting.

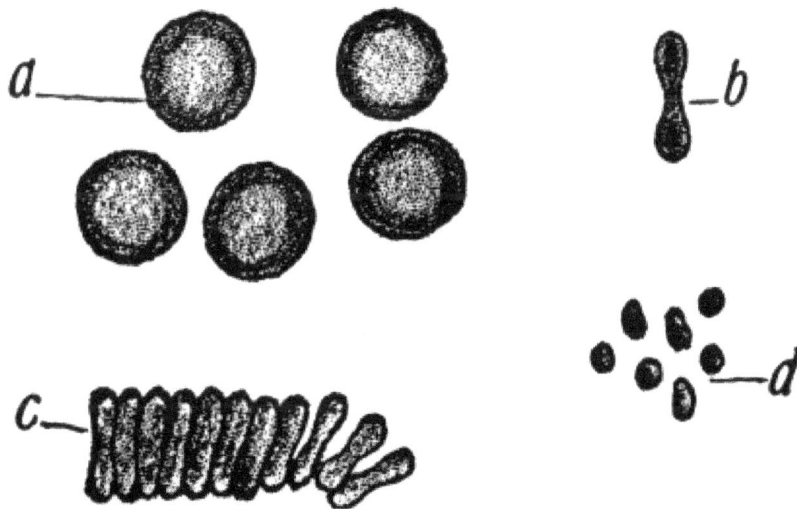

Fig. 45.—Cells of blood: *a*, Colored blood-corpuscles seen on the flat; *b*, on edge; *c*, in rouleau; *d*, blood platelets. (Leroy.)

Blood-corpuscles.—The solid parts of the blood are the red corpuscles, the white corpuscles, and the blood plaques or plates. It is to the *red* _corpuscles_, or _erythrocytes_ which number about 5,000,000 to the cubic millimeter of blood, that the color of the blood is due. Under the microscope they appear as small, spherical, biconcave discs with a slightly greenish-yellow color, which have a tendency to form in rouleaux. They are homogeneous, with no limiting membrane, and are made up of a fine network of tissue, the _stroma,_ in which is embedded the hemoglobin or coloring matter. This hemoglobin is a crystalline body and the most complex substance known to chemists. The corpuscles are very flexible and can squeeze through small apertures, as in the tiny capillaries, and regain their shape. They are probably formed chiefly in the red bone marrow at the ends of the bones, which under the microscope shows red corpuscles in various stages of growth, and also in the spleen, for which no other use is known. Their function is to carry oxygen, which forms a chemical combination, though an extremely loose one, with the hemoglobin. As the tissues are more greedy of oxygen than is the hemoglobin, they rob the corpuscles of it.

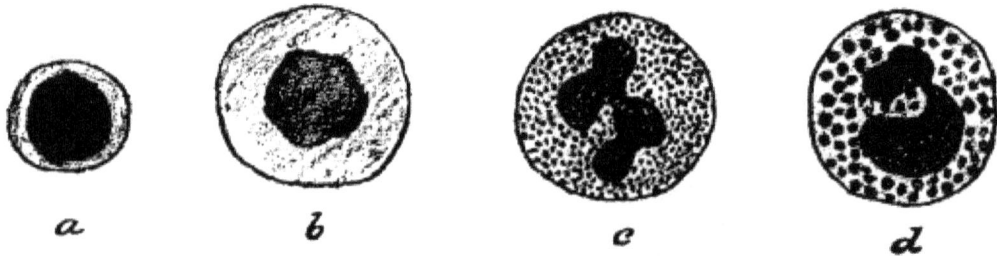

FIG. 46.—Various forms of leucocytes: _a,_ Small lymphocyte; _b,_ large lymphocyte; _c,_ polymorphonuclear neutrophile; _d,_ eosinophile. (Leroy.)

The _white corpuscles_ or _leucocytes_ are much fewer in number, about one to from 300 to 700 of the red, the average number being 5,000 to 10,000 to the cubic millimeter. They are larger than the red corpuscles, colorless, and spherical when at rest. Their structure is more definite, there being a definite cell substance or protoplasm and one or more nuclei, which vary more or less in shape and size. The corpuscles are classed in accordance with these variations in the nuclei. They are most numerous during digestion and are probably formed in the lymphatic system, constantly passing from the lymphatics to the arteries and veins. For they have the function of amœboid movement by which they not only wander from place to place in the blood, keeping close to the sides of the vessels, but pass through the walls of the capillaries, probably between the cells which form their lining, into the lymph spaces. This is known as *migration of the white corpuscles.* In inflammation they collect in the inflamed area to assist in allaying the inflammation by absorbing and carrying off its products. For they carry waste products and destroy poisons, acting as scavengers and protectors of the body. When they are unsuccessful and the inflammation gets the better of them, they become *pus corpuscles.*

Besides the corpuscles there are seen floating in the blood small disk-like substances with no special characteristics, the *blood plaques* or *plates*, whose function is unknown.

In *anemia* the red corpuscles are diminished and the white corpuscles and blood plaques increased in number. After excessive bleeding normal salt solution is injected, subcutaneously or by rectum, as being nearly equivalent to blood serum in composition, and the renewal of the solid elements is left to time. The length of time needed for their restoration is about a week, except in the case of the hemoglobin, which takes longer.

CHAPTER IX.
THE LUNGS AND RESPIRATION.

Besides the heart and the great vessels the chest contains the lungs, the chief organ of respiration, which, with the rest of the respiratory system, will now be treated. The nose and mouth, through which the air first enters the body, have already been spoken of. From them the air passes through the larynx to the trachea, thence to the bronchi, and so to the lungs, where the supply of oxygen for the tissues is taken from the air by the hemoglobin of the blood.

The Larynx.—The larynx lies in front of the pharynx at the upper and fore part of the neck, where it causes a considerable projection, known as *Adam's apple*. It is a triangular box, base up, flattened at the back, in front, and at the sides, but becoming cylindrical below. Above it opens into the bottom of the pharynx and below into the trachea. It is lined with mucous membrane. Its opening at the base of the tongue is closed during swallowing by a little door-like valve of fibro-cartilage, the *epiglottis*, to prevent the entrance of food.

Nine *cartilages* go to make up the larynx, of which the most important are the thyroid and cricoid cartilages and the epiglottis already mentioned. The *thyroid* is the largest and is open behind, its two alæ or wings meeting in an acute angle in front and forming the Adam's apple, always more prominent in the male than in the female. It is attached above to the hyoid bone and has cornua or horns on either side, top and bottom. The *cricoid* or ring-like cartilage resembles a seal ring with the stone placed posteriorly. It is stronger than the thyroid and forms the lower part of the cavity of the larynx. Inside and resting on the upper border of the cricoid are the two smaller *arytenoid* or pitcher-like cartilages, pyramidal in shape, and surmounting these again the two *cornicula laryngis*. The two *cuneiform cartilages* are in the free borders of the folds of mucous membrane which extend from the apex of the arytenoids to the sides of the epiglottis. Numerous small muscles serve to bind these various cartilages together.

At the angle of the thyroid cartilage in front are attached the epiglottis at the top and just below that the *superior* or *false vocal cords*, two folds of mucous membrane enclosing the *superior thyro-arytenoid ligaments*. Lower still are found the *inferior* or *true vocal cords*, which are formed by the *inferior thyro-arytenoid ligaments* covered with a thin, tightly fitting mucous membrane. Both sets of vocal cords as well as the epiglottis may be seen by means of a head and a throat mirror. Between the true vocal cords is a narrow triangular interval called the *glottis*. It is by means of the vibrations of these cords that sound is produced. The false vocal cords cannot produce sound, though they can modify it indirectly. Quality of voice, as treble, base, etc., depends upon the size of the larynx and the length and elasticity of the vocal cords. Modulation is produced by changing the form of the cavity of the mouth and nose. In whispering the lips take the place of the vocal cords and produce sound by the vibration of their muscular walls.

Instead of tracheotomy *laryngotomy* is sometimes done in the depression between the thyroid and the cricoid, which may be felt on the living. *Foreign bodies* sometimes get into the larynx and have to be removed, or the mucous membrane may become inflamed, causing *laryngitis*. *Syphilis* attacks the larynx, and *tuberculosis* and *cancer* of the larynx occur, these last two being generally fatal. *Edema of the glottis* may also occur.

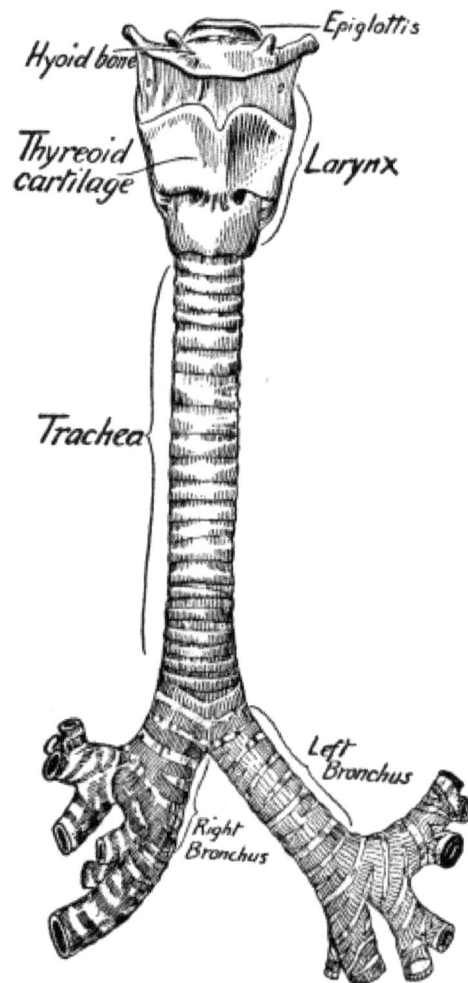

FIG. 47.—The larynx, trachea and bronchi
(After Sobotta.)

The Trachea.—The trachea is a membranous tube extending down from the larynx for about four and a half inches to the fourth or fifth dorsal vertebra, where it divides into the right and left bronchi. It is formed of sixteen to twenty imperfect cartilaginous rings, open behind, enclosed in a double elastic fibrous membrane, and is lined with ciliated mucous membrane. The rings are for strength and in the interval at the back where they are wanting there is one layer of longitudinal and another of transverse unstriped muscle fibers. The passage is kept clear by the action of the cilia, which sweep up and out any particles of dust that become entangled in the mucus. *Tracheotomy* is generally done about one inch below the cricoid, just above the sternal notch, incision being made through the cartilage.

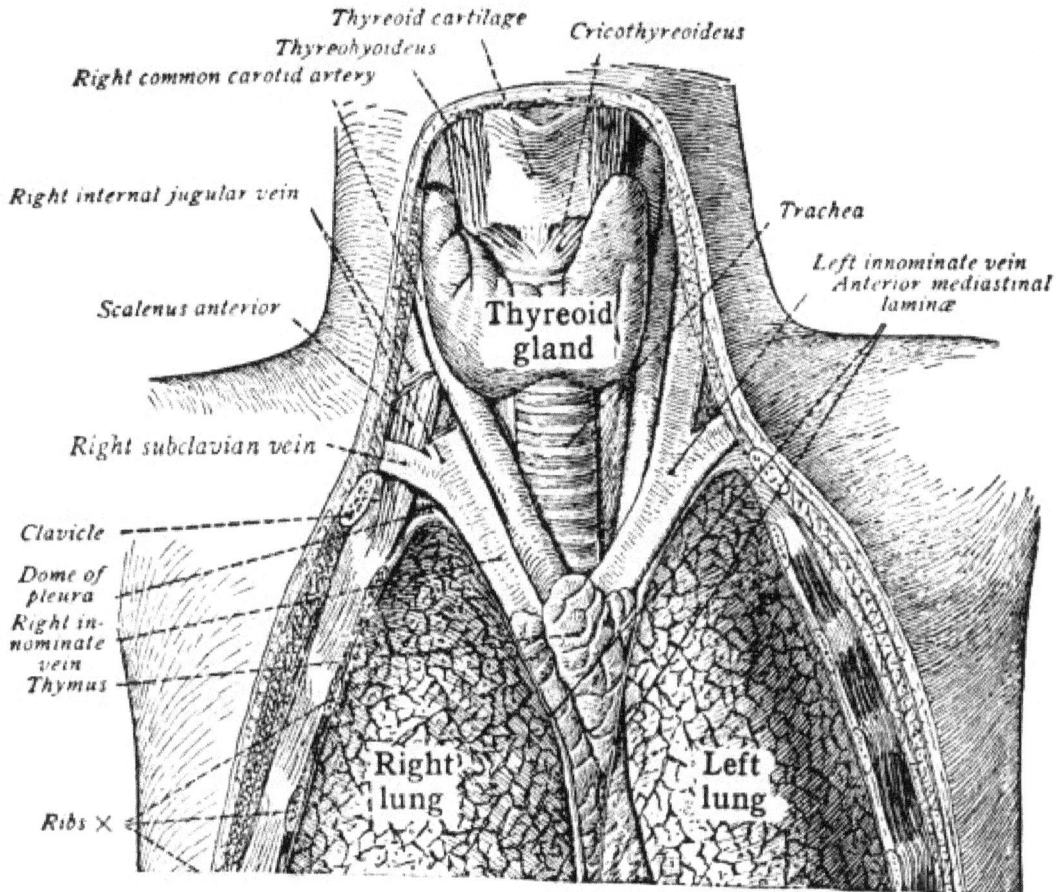

Fig. 48.—The upper thorax of a child eight years old, showing the thyroid and thymus glands. (Sobotta.)

Extending up on either side of the upper trachea in the neck are the two lobes of the *thyroid gland*, the isthmus, which connects the lobes, covering the trachea below anteriorly. The function of the gland is obscure, but it has an internal secretion of great importance in the metabolic processes. Its removal or disease is followed by general disturbances of mind and body. The injection of thyroid extract has proved effective as treatment. In *goiter* the gland becomes enlarged. The *thymus gland* lies below the thyroid gland at birth, in front of and at the sides of the trachea, and runs down behind the sternum. It is largest at the end of the second year, after which it atrophies, being almost absent at puberty. It, too, is ductless and its function is not well understood.

The Bronchi.—The two bronchi, of which the right is the larger and shorter, resemble the trachea in structure. As they enter the root of the lung they divide, the right into three and the left into two branches, one for each lobe, after which they divide and subdivide, the bronchioles becoming smaller and smaller and finally ending in the *infundibula*, pouch-like places lined with air cells, in which cilia keep the air in motion. As they grow smaller the bronchioles gradually become wholly membranous.

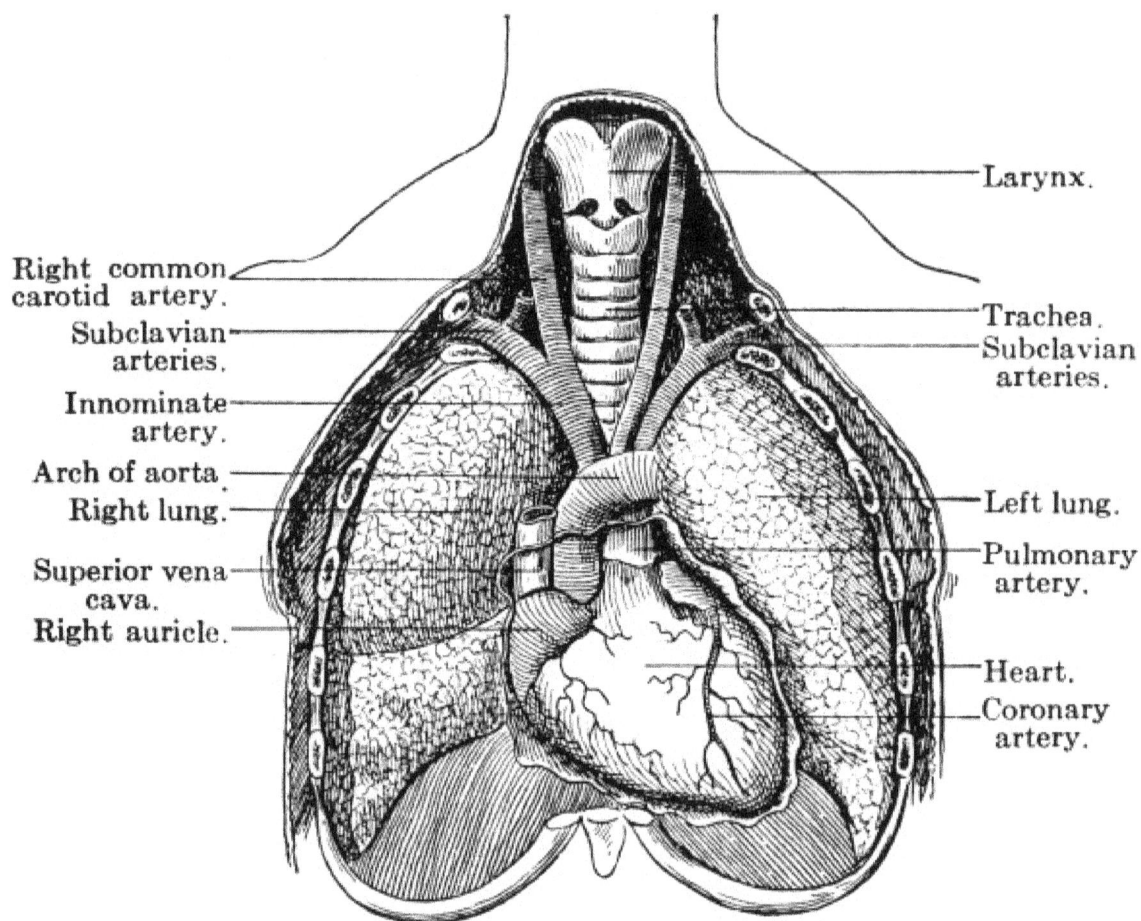

Fig. 49.—Relation of lungs to other thoracic organs.
(Ingals.)

The Lungs.—The lungs themselves, two in number, lie each in a serous sac or *pleura*, similar in structure to the pericardium and serving a like purpose. The outer layer of the pleura is reflected back over the thoracic wall and diaphragm. There is no pleural cavity in health between the two layers of the pleura, the two surfaces being in close contact, though moistened with lymph to prevent friction during respiration. In inflammation of the pleuræ or pleurisy, they become thickened and roughened and friction results, as is shown by the sounds heard through the stethoscope. Friction causes effusion and fluid collects. This generally absorbs again, but occasionally the serous fluid becomes pustular and *empyema* results.

In front, between the two pleuræ, which are wholly separate, is the *mediastinal space* or *mediastinum,* which extends from the sternum to the spinal column and contains all the thoracic viscera except the lungs and heart, that is, the trachea, esophagus, thoracic duct, and many large vessels and nerves.

Fɪɢ. 50.—Diagrammatic representation of the termination of a bronchial tube in a group of infundibula: *B*, Bronchial tube; *LB*, bronchiole; *A*, atrium; *I*, infundibulum; *C*, alveoli. (de Nancrede.)

Roughly speaking, the lungs begin at the sterno-clavicular articulation above, the apex coming up above the level of the first rib, and extend downward together to the fourth cartilage, where the lower margins gradually separate, the lowest lung limit being the eleventh rib in the vertebral region. Each lung is conical. The apices extend upward and the bases, which are broad and concave, rest upon the diaphragm. The right lung is divided by a fissure into three lobes, the left into two. The root consists of a bronchus and pulmonary arteries, veins, lymphatics, and nerves. The tissue itself is composed of an aggregation of lobules, each consisting of a terminal bronchiole with its alveoli or air cells, blood-vessels, and nerves, a lung in miniature. The blind pouches which the air cells surround are called *infundibula* and are separated by delicate membranous septa in which lie the capillaries of the pulmonary artery, thus exposing the blood to the air on two sides. The lung itself is supplied by the bronchial arteries from the thoracic aorta and by branches of the sympathetic and pneumogastric nerves.

At birth the lungs are pinkish-white in color but in later life they are marked with slate-colored patches, due to the deposit in the lung tissue of particles of dirt breathed in. They are light, spongy, and highly elastic, and will float in water, crepitating upon pressure owing to the air in the tissue.

At birth, also, the lungs are solid, so that the first air has to overcome adhesions between the collapsed walls of the bronchioles and air sacs, but after they are thus gradually unfolded, in that they are of extensible material and open to the air above, atmospheric pressure from within keeps them distended to the full extent of the chest, which is air tight. They never collapse afterwards unless puncture of the chest wall, as in stabbing, causes collapse, in which case the lung shrivels into a small ball.

Respiration.—That the organic materials used by the body as food may give up their energy they must be broken up, and for this oxygen is needed. The supply of oxygen for the purpose is brought to the tissues by the blood, which acquires it in the lungs, and the waste product of combustion, carbon dioxide, is carried off in the same manner. The lungs are, therefore, adapted to take in large quantities of air and to keep up a rapid exchange of oxygen and carbon dioxide in the blood. This process of supplying oxygen to the tissues and of removing carbon dioxide and other waste is ordinarily an involuntary act, though it can be regulated temporarily, and is known as respiration or breathing.

There are two periods to respiration: 1. *inspiration* or the drawing in of air, and 2. *expiration* or the expulsion of air from the lungs, the former process being a little shorter than the latter. A pause follows each expiration before there is another inspiration. At birth the normal rate of respiration is 42, but it grows slower as the child grows older, being 26 at the age of five or six, while in the adult it averages 17 to 20 times a minute. It is slower during sleep and more rapid during physical activity. The average amount of air taken in with every inspiration is 30 cubic inches and the minimum air space per individual should be 3000 cubic feet per hour.

Breathing is of two kinds, *diaphragmatic* or *abdominal* and *chest* or *rib* breathing, the former usually being more pronounced in men than in women, probably because of centuries of tight dressing on the part of the latter. As a rule, however, both diaphragm and ribs come into play; for in inspiration, which is an active movement, the thorax becomes enlarged from before backward, laterally, and vertically. The ribs are raised by the external intercostals chiefly, though the internal intercostals aid somewhat, and swinging out upon the vertebræ, widen the chest as well as deepen it. The diaphragm, which is dome-like when relaxed, becomes flattened in contraction and so increases the size of the chest from above downward. As the chest enlarges, the lungs expand, the air in them becomes rarefied, and more air rushes in. When the lungs are full they relax and the muscles relax after their contraction, so that expiration is a passive movement, due largely to the elastic relaxation of lungs and muscles, the air being driven out by the lessened capacity of the lungs.

Difficult Breathing.—In heart and lung troubles, where too little oxygen is carried to the tissues, dyspnœa or difficult breathing results and may even advance to *asphyxia,* a condition in which no air is obtained. In difficult or labored respiration the pectoral muscles are used in inspiration and the scaleni, which pass from the vertebræ of the neck to the sternum, develop and become powerful. The levatores of the ribs may also assist, and even the muscles of the neck and arms may help out, while in forced expiration the abdominal muscles are called into play. The glottis opens and closes rhythmically as the air enters and leaves the lungs, and the nostrils add their mite in the struggle for oxygen. Finally there may be scarcely a muscle in the body that is not striving to aid the respiration, and general convulsions may result, followed by exhaustion and death.

Air.—In ordinary breathing the lungs are not used to their full capacity and the air ordinarily used is known as *tidal air.* In forced inspiration the lungs are filled to their fullest extent and the air then taken in in excess of the tidal air is known as *complemental air.* In like manner, the difference between the air ordinarily breathed out and that breathed out in forced expiration is known as *supplemental air.* The sum of these three is the *vital capacity* of the lungs, while beyond this there is probably some air that is never expelled, the stationary or *residual air.*

Respiratory Sounds.—The entrance and exit of the air is accompanied by respiratory sounds or murmurs, which vary according to their position in the trachea, the bronchi, or the bronchioles and are modified in diseases of the lungs and bronchi, when they are often called *râles.*

Changes in Air in Lungs.—In passing through the nose and the rest of the respiratory tract the air is warmed to body temperature and saturated with moisture. After its entrance into the lungs various changes take place in it through the mingling of the tidal with the residual air. Thus, it gives up about 4 or 5 per cent. of its oxygen and acquires some 4 per cent. additional carbon dioxide, while the amount of nitrogen remains about the same. By its giving up more oxygen than it receives carbon dioxide, its volume is slightly diminished. Exhaled air also contains traces of ammonia and certain organic matters, generally the results of decomposition, which give a bad odor to the breath and are more dangerous in a close room than the mere lack of oxygen or the presence of carbon dioxide. Indeed, the amount of oxygen may be very much diminished, being reduced even to 5 or 6 per cent. instead of the normal 21 per cent., without being noticed or giving rise to any immediate bad results. Yet the importance of ventilation is very evident.

Effect on Blood.—Respiration causes changes also in the blood, the venous blood being purple and the arterial bright red. This difference in color is due to the absence or presence of oxygen, which is not absorbed or dissolved by the blood but forms a rather unstable compound, *oxyhemoglobin*, with the hemoglobin of the blood. As the oxygen is removed in the passage of the blood through the body, there results in venous blood reduced hemoglobin, which is of a purplish color. Upon exposure to the air, however, it absorbs oxygen once more and resumes its scarlet color. If carbon monoxide gets into the blood, as in cases of gas poisoning, it drives off the oxygen and forms a more stable compound with the hemoglobin, whence the difficulty in restoring a person so poisoned.

Nervous Mechanism.—Nervously, respiration is controlled in three ways: 1. by the phrenic nerve to the diaphragm; 2. by some fibers of the vagus or pneumogastric, and 3. by the respiratory center in the bulbous portion of the spinal cord. Injury to the respiratory center means the ceasing of respiration and death. Stimulation of the respiratory center seems to depend upon the character of the blood. If it is well oxygenized, the breathing is slow and quiet; if there is a lack of oxygen, dyspnœa results. Probably certain chemical substances in the blood, which are ordinarily rapidly burned up by the oxygen but which accumulate in its absence, serve to stimulate the respiratory center, thus adjusting the effort to get oxygen to the need of it. Respiration may be stopped by stimulating the mucous membrane of the nose, as with strong ammonia.

Variations.—Certain variations from the ordinary respiration might be mentioned here. A deep inspiration followed by a long expiration is known as a *sigh* and a very deep inspiration through the mouth only as a *yawn*. *Hiccough* results from a sudden inspiratory contraction of the diaphragm during which the glottis is suddenly closed. In *sobbing* the inspirations are short and rapid with a prompt closing of the glottis between. Both *coughing* and *sneezing* consist of a deep inspiration followed by complete closure of the glottis and then its sudden opening and the forcible expulsion of air. Coughing, however, is generally caused by an irritation or obstruction of the larynx or trachea and the air is expelled through the mouth, while sneezing is caused by irritation of the nasal passages and the air is driven out through the nose. *Laughing* and *crying* also resemble one another in that each is an inspiration followed by a series of short, spasmodic expirations, during which the glottis is open and the vocal cords in characteristic vibration. They differ, however, in rhythm and in the facial expression that accompanies them.

CHAPTER X.
THE ABDOMEN AND THE ORGANS OF DIGESTION AND EXCRETION.

The Abdominal Cavity.—Below the diaphragm and separated from the lowest cavity of the trunk, the pelvis, only by an invisible plane drawn through the brim of the true pelvis, is the abdominal cavity, which may be said in a general way to contain the organs of digestion and the kidneys. It is protected behind by the vertebræ and anteriorly by the lower ribs above and below by muscular walls, which make possible the complete bending of the body. These muscles are for the most part large and very strong and the greater number are inserted, in part at least, into a median tendinous line, the *linea alba*, which passes from the ensiform cartilage of the sternum above to the symphysis pubis below.

Muscles.—The *external oblique muscles* form the outermost layer of the abdominal wall. They rise from the external surface of the eight lower ribs on either side and are inserted in the anterior half of the iliac crest as well as by aponeurosis in the linea alba, where each joins its fellow from the opposite side, the fibers running downward and inward like the fingers in the trouser's pocket. Along the lower border of the aponeurosis is a broad fold, *Poupart's ligament*. The *internal oblique* rises on either side from the outer half of Poupart's ligament and the anterior part of the crest of the ilium and is inserted into the crest of the os pubis, the cartilages of the lower ribs, and the linea alba. Its fibers run at right angles to those of the external oblique. These oblique muscles serve to compress the viscera, to flex the body, and also assist in expiration.

The deepest of the abdominal muscles is the *transversalis*, which rises from the outer third of Poupart's ligament and the adjoining part of the crest of the ilium, from the six lower costal cartilages, and by a broad aponeurosis, the *lumbar fascia*, from the lumbar vertebræ. It is inserted into the pubic crest and by aponeurosis into the linea alba. There is one of these muscles on either side.

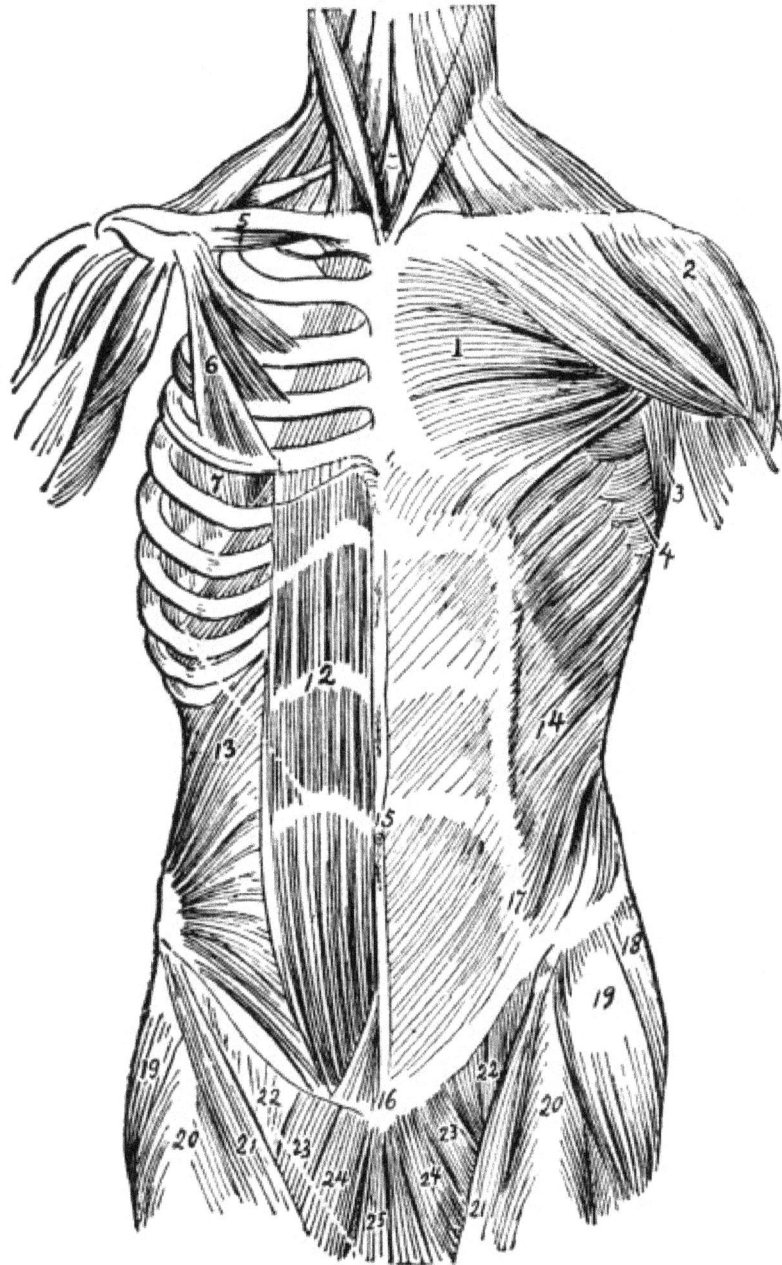

FIG. 51.—Muscles of the trunk from before (left side, superficial; and right side, deep): 1, Pectoralis major; 2, deltoid; 3, portion of latissimus dorsi; 4, serratus magnus; 5, subclavius; 6, the pectoralis, sternocostal portion; 7, serratus magnus; 12, rectus abdominis; 13, internal oblique; 14, external oblique; 15, abdominal aponeurosis and tendinous intersections of rectus abdominis; 16, over symphysis pubis; 17, linea semilunaris; 18, gluteus medius; 19, tensor vaginæ femoris; 20, rectus femoris; 21, sartorius; 22, femoral part of iliopsoas; 23, pectineus; 24, adductor longus; 25, gracilis. (Dorland's Dictionary.)

The *rectus abdominis* is also really two muscles and extends from the symphysis pubis to the cartilages of the fifth, sixth, and seventh ribs. At first it passes back of the oblique and transversalis muscles, but about a fourth of the way up it passes in front of the transversalis and between two layers of the internal oblique, which thereafter forms its sheath. Its chief duty is to flex the chest on the pelvis, though it also compresses the abdominal viscera.

One other muscle, a small one, is found in front, the *pyramidalis*, which rises from the pubic crest and is inserted into the linea alba midway to the umbilicus.

At the back the open space over the kidneys, between the lower ribs and the os innominatum, is closed in on either side by the *quadratus lumborum*, which extends from the three or four lower lumbar vertebræ and the adjacent iliac crest to the last rib and the upper four lumbar vertebræ. It flexes the trunk laterally or forward according as one muscle or both are used, and may aid in either expiration or inspiration.

The *nerves* of the abdominal muscles are chiefly the internal intercostals.

The Peritoneum.—Lining the abdominal cavity is a serous membrane, the peritoneum, which is reflected back over the viscera within in such a way as to cover each one wholly or in part. Folds of peritoneum, the *omenta*, connect the stomach with the other viscera, the most important being the *great omentum*, which has one layer descending from the anterior and another from the posterior wall of the stomach. The *mesenteries* are double layers of peritoneum which hold the intestines to the vertebræ and posterior wall. Between their folds run the blood-vessels.

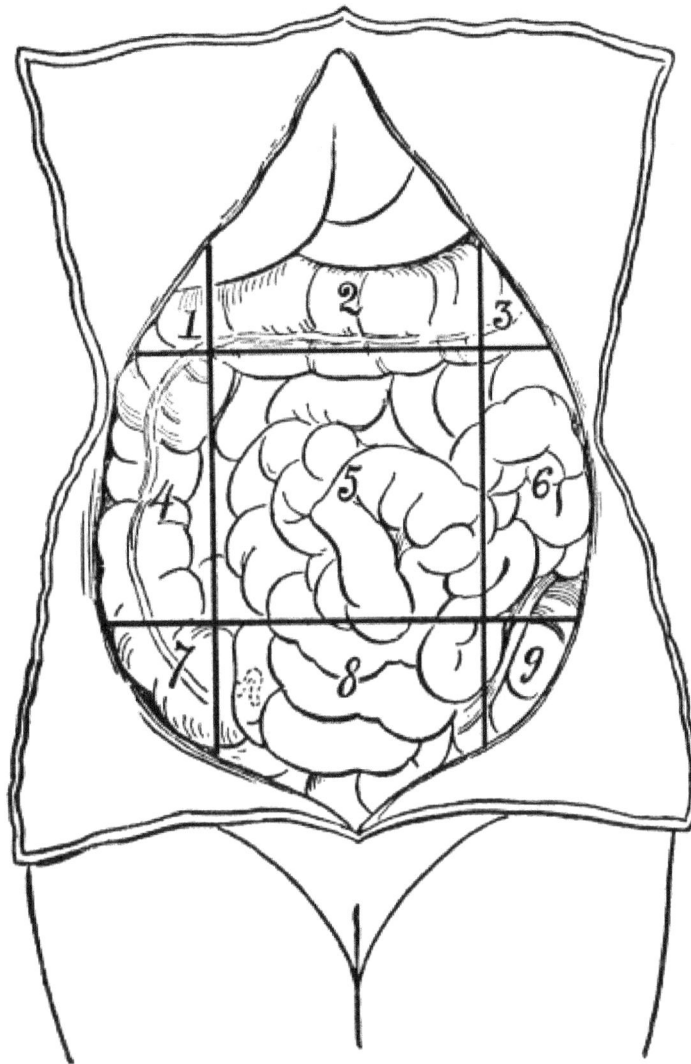

FIG. 52.—Diagram showing the nine regions of the abdominal cavity: 1, Right hypochondriac; 2, epigastric; 3, left hypochondriac; 4, right lumbar; 5, umbilical; 6, left lumbar; 7, right iliac; 8, hypogastric; 9, left iliac. (Ashton.)

Abdominal Regions.—For convenience of description the abdominal cavity has been divided into nine regions by means of two transverse parallel lines, the one through the ninth costal cartilages and the other just over the iliac crests, and two perpendicular parallel lines through the cartilage of the eighth rib and the middle of Poupart's ligament on either side. These nine regions have been named as follows: The right and left hypochondriac regions up under the ribs with the epigastrium between, the right and left lumbar regions next below with the umbilical between, and the right and left inguinal with the hypogastric between. Others divide it into quadrants by one line drawn across and another down through the umbilicus. The contents of the abdomen in full are the stomach, intestines, liver, gall-bladder, spleen, pancreas, kidneys, suprarenal capsules, and the great vessels, that is, the organs of digestion and excretion. When distended the bladder extends up into the abdominal cavity, as does the uterus also when enlarged.

Salivary Digestion.—Although most of the digestive organs are situated in the abdomen, the food enters the body through the mouth, where its prehension is a voluntary act. Here digestion also begins and from the first the process is a double one, mechanical and chemical, mechanical digestion consisting largely of muscular movements by which the food is ground up and carried through the digestive tract. Thorough mastication or grinding of the food by the teeth is necessary, while the tongue assists by moving the food about and by mixing it thoroughly with the *saliva*, a viscid fluid composed of water and salts and having a slightly alkaline reaction. The saliva is secreted by the parotid, sublingual, and submaxillary glands, and serves to soften and dissolve the food and by virtue of its unorganized ferment, *ptyalin*, to convert starch into sugar. Upon proteins and fats it has practically no digestive action. Moderate warmth and an alkaline medium favor its action, while extremes of heat or cold or an acid medium hinder it. There is little absorption in the mouth, though starch, nicotine, and alcohol may be absorbed in small quantities.

The Pharynx.—When the food is ready for deglutition or swallowing, it is thrust back into the pharynx, a somewhat conical, musculo-membranous sac, situated, base upward, behind the nose and mouth and behind, but somewhat above, the larynx. The pharynx is about four and a half inches long and ends on a level with the cricoid cartilage in the esophagus or gullet. It is attached to the vertebræ at the back and opens in front into the mouth. The posterior nares, the Eustachian tubes, and the larynx also open into it, the last being protected by the *epiglottis*, which closes during deglutition to prevent food from entering the air passages, just as the soft palate is drawn back to prevent regurgitation of food into the nose. There are three coats to the pharynx: 1. a mucous coat continuous with that of the mouth and ciliated down to the floor of the nares; 2. a fibrous coat, and 3. a muscular coat containing among others the constrictor muscles which serve to carry the food down to the esophagus. Its arteries are branches of the external carotid and its nerves come from the spinal accessory and the sympathetic. Occasionally a *foreign body* gets lodged in the pharynx just out of reach of the finger and threatens strangulation. *Retropharyngeal abscess* on the posterior wall occurs rarely.

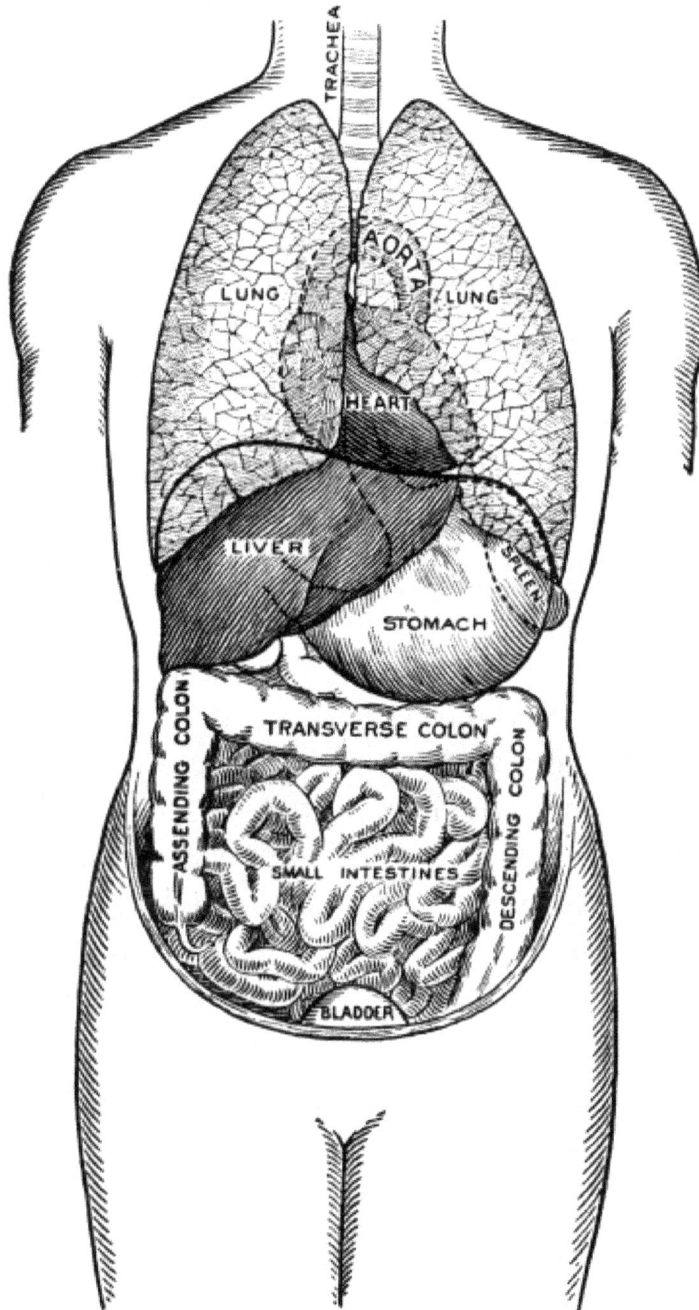

FIG. 53.—Position of the thoracic and abdominal organs, front view. (Morrow.)

The Esophagus.—From the pharynx the food passes to the cardiac orifice of the stomach, opposite the tenth dorsal vertebra, through the esophagus, a muscular tube about nine inches long, which collapses when empty, its lumen then appearing as a transverse slit. It, too, has three coats: 1. an inner mucous coat; 2. an areolar coat, and 3. a muscular coat, the muscles being arranged in two sets, an outer longitudinal layer and an inner circular layer. By a series of rhythmic contractions, especially of the circular fibers, the food is pushed along, though sometimes with liquid food there is no peristaltic action of the esophagus, the pharyngeal muscles alone sending it to the stomach. At the lower end of the esophagus an especially strong band of circular muscle fibers form a sort of sphincter, which prevents the regurgitation of food. The whole act of swallowing is a reflex, not a voluntary act and is due to irritation set up by the stimulus of the foreign body, the food. *Stricture of the esophagus* is common and may be of three kinds: 1. spasmodic, occurring in nervous women; 2. fibrous, due to scar tissue, or 3. malignant, due to cancer.

The Stomach.—The stomach is a pear-shaped dilatation of the alimentary canal, lying under the liver and diaphragm in the epigastrium and left hypochondrium and connecting the esophagus with the small intestine. It lies largely behind the ribs, but the greater curvature is only two fingers' breadth above the umbilicus and can be manipulated through the skin. The *cardiac end*, into which the esophagus enters, is the larger and points upward to the left. The lesser and lower end, known as the *pylorus*, is at the right and its opening into the small intestine is guarded by the *pyloric sphincter*. The lesser curvature is concave and on the upper surface; the greater, convex and on the under surface. The great omentum is attached to the latter.

In *size* the stomach varies more or less, that of a man generally being larger than that of a woman, but it is usually about ten inches long and four or five inches across. It has a capacity of about five pints and serves as a storehouse for food.

The stomach has four *coats*: 1. a serous coat derived from the peritoneum; 2. a muscular coat of three layers with longitudinal fibers continuous with those of the esophagus, circular fibers, and oblique fibers; 3. an areolar coat, and 4. a mucous coat, which, when the stomach is empty, is thrown into longitudinal folds or *rugæ*, and whose surface is covered with glands, the *gastric glands*, for the secretion of the digestive fluids.

The *arteries* come from the celiac axis and the *nerves* from the pneumogastric and the solar plexus.

Ulcer and *cancer* of the stomach are both rather common. In the former there is apt to be *hyper*-acidity and in the latter *hypo*-acidity, but the rule does not always hold. In cases of ulcer there may be hemorrhage and even perforation. Such hemorrhage can be distinguished from hemorrhage from the lungs by its slightly acid odor and by the frothy character of hemorrhage from the lungs. There is much irritation at the pylorus and where there is irritation there is liable to be cancer.

Gastric Digestion.—In the stomach the food is churned and thoroughly mixed with the gastric juices, and it is also subjected to a propulsive movement that drives it on to the intestine. When it comes to the stomach it is semi-solid and when it has become fluid or semi-fluid, in which state it is known as *chyme*, it is ready to pass on. Before it can do so, however, it must overcome the strong pyloric sphincter, and this it does by the muscles about the sphincter pushing it constantly on until the sphincter gives way. Probably most of the propulsive movements take place within a few inches of the pylorus.

The *gastric juice* is secreted by glands in the wall of the stomach and poured out through little tubules which project from the surface. It is a thin, almost colorless fluid with a sour taste and odor due to the presence of free *hydrochloric acid*, an important element in digestion. Probably when the stomach is empty and for some twenty minutes after the appearance of food there is no hydrochloric acid present and, the food being alkaline, salivary digestion continues. Then, called forth by the presence of the food, the hydrochloric acid appears and salivary digestion ceases in the acid medium. Little digestion of starches or fats takes place, the chief action being on proteins, which are converted into soluble peptones. For besides hydrochloric acid the gastric juice contains two ferments: 1. *pepsin*, which is particularly active in aiding the digestion of proteins, and 2. *rennin*, which especially affects milk. Neither hydrochloric acid nor pepsin seems capable of digesting food alone, but each is essential to the other. They are secreted by different types of cells, secretion depending upon the nerve supply and upon the presence of food. Gastric digestion is favored by minute subdivision of the food and by the right proportion of hydrochloric acid, which should be 0.2 per cent. Body temperature is also advantageous. Except that proteins are put in solution and partly digested, little digestion goes on in the stomach, and though the rugæ afford a large absorbing surface, little absorption takes place, although more takes place than in the mouth and in time most foods, except fats, can be absorbed. The time of digestion varies with different foods and in different people, but probably three to five hours are necessary. The food leaves the stomach as *chyme*, a fluid of about the consistency of pea soup.

Vomiting is more or less the reverse of swallowing and is generally preceded by a feeling of *nausea*, which starts up *retching*, a more or less involuntary effort of the stomach to throw off its contents. To relieve the retching a long breath is taken, followed by a deep expiration that opens the cardiac end of the stomach and allows the abdominal muscles to force the food out. After much vomiting and prolonged retching the pyloric end of the stomach may be affected and bile will then appear in the vomitus. *Artificial vomiting* may be produced by irritation of the gastric nerve center in the brain or by irritation of the stomach itself.

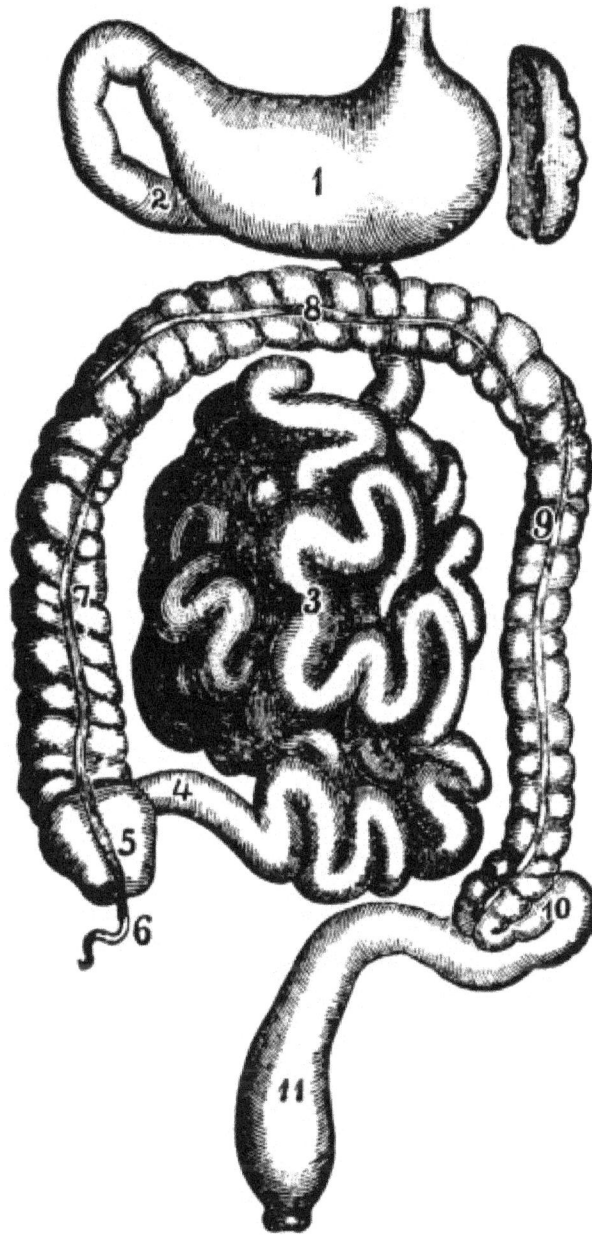

FIG. 54.—The intestinal canal: 1, Stomach; 2, duodenum; 3, jejunum; 4, ileum; 5, cecum; 6, vermiform appendix; 7, ascending colon; 8, transverse colon; 9, descending colon; 10, sigmoid flexure; 11, rectum. (Leidy.)

Intestinal Canal.—From the stomach the food passes into the intestinal canal, a convoluted tube which extends from the stomach to the anus and in which, more particularly in the upper portion, the greater part of the digestion and absorption of food takes place. This tube, which is about six times the height of its possessor, consists of two parts, the small and the large intestines, the first four-fifths, or about 25 feet, being small intestine.

It occupies the central and lower parts of the abdominal cavity and a small portion of the pelvic cavity, and is attached to the spine by the mesentery, which, however, allows great freedom of motion, so that there is little fixation to the loops of the small intestines.

The Small Intestine.—The small intestine opens out of the stomach and has three divisions: 1. the _duodenum_, which is only about ten to twelve inches long; 2. the _jejunum_, so called because it is generally empty after death, which is about two-fifths of the remainder and lies chiefly in the umbilical region and the left iliac fossa, and 3. the _ileum_ or _curved intestine_, the remaining three-fifths, which gets its name from its numerous coils and which lies in the middle and the right side of the abdomen. There is no direct division between the jejunum and the ileum, but the first part of the former and the last part of the latter are quite different in character. At its entrance into the large intestine the ileum is guarded by the _ileo-cecal valve_.

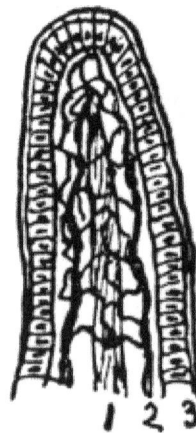

FIG. 55.—
1, Central lacteal;
2, capillary network;
3, columnar cells.

The same _coats_ continue in the small intestine as were found in the stomach, but they are here much thinner and the inner coat is shaggy, like velvet, with innumerable minute processes called _villi_, which greatly increase the absorbing surface. In fact, the great length of the intestine as well as the presence of the villi is aimed to provide a large surface to absorb the food as it passes, an even greater increase of surface being provided by the fact that the intestinal wall is thrown into folds, the *valvulæ conniventes*. Each villus is covered with a layer of columnar epithelial cells and has within connective tissue, in which are found a fine capillary network and open lymph spaces from which leads a single lacteal vessel.

Closely connected with the lymphatic vessels are the *solitary glands*, small round bodies the size of a small pin's head. *Peyer's glands* or *patches* are patches of solitary glands opposite the mesenteric attachment and are largest and most numerous in the ileum. In typhoid fever they are involved and may become the seat of ulcers. There are also the *glands of Lieberkühn* which secrete the succus entericus.

The *arteries* of the small intestine, which include the superior mesenteric, are from the celiac axis and the *nerves* are from the superior mesenteric plexus of the sympathetic. The *veins* empty chiefly into the portal system.

The *movements* of the intestine, like those of the esophagus, are peristaltic, but the action is complicated by the fact that the tube is not straight but in coils.

Intestinal Digestion.—The food, which enters the duodenum as chyme, there comes in contact with the bile and the pancreatic juice, which together but unmixed enter the duodenum from their respective ducts by a common orifice. As in the stomach, the digestive juices are called forth by the presence of food. The *bile* is secreted in the liver, from which it flows away through the hepatic duct, which joins the cystic duct from the gall-bladder to form the common bile duct. Through this it flows into the intestine during digestion, but between whiles it passes up into the gall-bladder, where it is stored for future use and whence it is expelled when needed. When pure it is a thick, viscid liquid, varying from a bright red to a greenish-yellow in color according to the pigments present, and of an alkaline reaction. It consists chiefly of the bile pigments, *biliverdin*, which gives the green color, and *bilirubin*, which gives the red color, and of bile salts in solution, *cholesterin*, which probably forms the basis of many gall stones, is also present. Bile is a disinfectant to the bowel and a lubricant for the feces. How much digestive action it has is a question, but it affords the necessary alkaline medium for the pancreatic juice to act in.

The *pancreatic juice* is secreted by the pancreas, from which it enters the intestine through the pancreatic duct, and is probably the most important fluid in the digestive process. It is clear, practically colorless, slightly viscid or gelatinous, and quite strongly alkaline in reaction, owing to the presence of sodium carbonate. It contains three ferments, *amylopsin* for the digestion of starch, *trypsin* for the digestion of proteins, and *steapsin* for the digestion of fats. By it, as by the saliva, starch is turned into sugar or maltose, in which form it is absorbed, while proteins are converted into peptones, as they are in the stomach. Since, however, fats are acted on nowhere else, the chief function of the pancreatic juice may be considered to be the digestion of fats. Having broken through their albuminous envelope, it divides them into glycerine and fatty acids and then emulsifies them with the assistance of the bile.

The food also comes in contact with the *succus entericus*, a juice secreted by the glands of Lieberkühn in the small intestine, whose chief action is the conversion of sugar into glucose.

Absorption.—As the food is absorbed from the intestine it is liquid and entirely digested and is known as *chyle.* Practically all absorption takes place from the small intestine, though there is a little in the large intestine. It takes place in two ways: 1. through the portal vessels and 2. through the *lacteals*, which are the lymphatic vessels of the small intestine. Fats are absorbed practically entirely by the lacteals. They enter the cells covering the villi, travel thence to the lymph spaces, and so into the lacteal or main lymph channel, whence they are carried to the thoracic duct and the general circulation. From the blood they are absorbed as fat and stored up as adipose or fatty tissue, which is found throughout the body in connective tissue about the organs. Organic salts and water are for the most part absorbed by the portal system, which they reach through the capillaries of the villi and through which they go to the liver. Starches, in the form of sugar, pass between the cells of the villi into the lymph spaces, from which they are taken up by the capillaries. On the way to the liver maltose becomes dextrose. Proteins, in the form of peptones, pass through the layer of epithelial cells to the lymph spaces and then to the capillaries, an active part being taken by the cells. By the time they reach the liver the peptones have been changed back into proteins. In fact, peptones seem to have some poisonous effect upon the blood if they get into it as such.

The Large Intestine.—The large intestine differs from the small in size and in fixity of position, lying curved in horseshoe shape above and around the small intestine. It is five or six feet long, large in caliber, and is thrown into crosswise folds. It has the same four *coats* as the small intestine, but the mucous coat is pale and smooth, without villi. Its *glands* are the crypts of Lieberkühn and the solitary glands. The *arteries* are branches of the superior and inferior mesenteric and the *nerves* come from sympathetic plexuses.

The blind sac lying in the right iliac fossa, with which the large intestine begins, is called the *cecum*, and into this the ileum opens, the ileo-cecal valve preventing regurgitation. Just below the ileo-cecal opening is the *vermiform appendix*, a narrow, worm-like tube with a blind end, varying in length from one to nine inches, but generally about four and one-half inches long, which, so far as is known, is functionless as well as dangerous. People have been born without an appendix and it has in rare instances grown again after operation. Its base is located in the living by *McBurney's point*, a point two inches from the anterior superior spine of the ilium on a line drawn from the spine to the umbilicus.

From the cecum the intestine ascends in what is known as the *ascending colon* along the abdominal wall at the right to the under surface of the liver, where it turns in the *hepatic flexure* abruptly across the body to the left, passing below the liver, stomach, and spleen in the *transverse colon*. In the *splenic flexure* it turns down the left abdominal wall, the *descending colon* passing to the crest of the ilium, where there is another curve, the *sigmoid flexure*, leading to the *rectum*, which passes for six or eight inches down along the vertebræ, a little to the left, to the *anus*, the external opening. This opening is guarded by two *sphincter muscles*, about an inch apart, the internal and external sphincters. The coils of the small intestine lie below the transverse colon, covered mostly by the omentum. The splenic flexure is behind the stomach and below the spleen and is slightly higher than the hepatic flexure. The sigmoid flexure can be felt in the *left inguinal region* in thin people.

The fact that the *rectum* is somewhat to the left is of importance in childbirth because if the rectum is packed, it may turn the child's head in the wrong direction.

No digestion goes on in the large intestine, the function being to dry by absorbing water. The *movements* are practically the same as those of the small intestine except that they are much less active. Fermentation makes the contents acid. By the time food reaches the rectum it has been thoroughly digested and has given up its nourishment. It is then expelled as waste matter or *feces*. *Defecation* combines the involuntary movements of peristalsis and relaxation of the sphincters with the voluntary aid of the abdominal muscles. The *ano-spinal reflex*, by which movements of the bowel are regulated, is in the lumbar enlargement of the cord.

The *hemorrhoidal veins* in the lower rectum are connected with both the systemic and the portal veins and have no valves so that, as they are subjected to much strain, they often become varicose and dilated. This condition is called *hemorrhoids* or *piles*. *Obstruction* of the intestine may be caused by the growth of a constricting band, by *intussusception* or telescoping of the intestine on itself, especially at the ileo-cecal valve, or by *volvulus* or twisting. *Foreign bodies* are sometimes found in the appendix but they are not usually the cause of appendicitis. *Cancer of the intestine* is common and its mass is apt to cause obstruction with all its attendant symptoms. It may necessitate an artificial anus. *Hernia* or *rupture* may also occur and the hernia may become strangulated.

Food and Metabolism.—Anything serves as food that replaces or hinders the loss to which the component parts of the body are liable. Proteins, carbohydrates, fats, some mineral matters, as salt and perhaps iron, and water are needed. The energy once expended by plants or animals in the formation of the materials which serve as food is set free in the body by the breaking up of these complex substances into their original elements, which are then recombined into the complex materials needed for the body's life and growth. This process of building up complex materials from simple ones is known as *anabolism* and that of breaking them down as *katabolism*, while the two combined form the complete cycle of *metabolism*. Those foods have the best value that give up their energy most readily. For their combustion, heat, oxygen, and water are needed. Hunger indicates that the supply of material for katabolism has been used up and that more is needed, just as thirst indicates the need of the system for more fluids.

The proteins or nitrogenous foods include all animal foods except fats, fish, crustaceans, eggs, milk and its products, certain vegetables, especially the lentils, that is, peas and beans, and gelatine. The fats include various fats and oils commonly eaten. The carbohydrates are the starchy foods, as cereals, sugars, fruits, and most vegetables, in fact, practically all except the lentils. Various beverages and condiments have no great nutritive value but serve to stimulate the appetite and to excite the secretion of the digestive juices. Coffee, tea, and alcohol are stimulants.

The different classes of foods have different functions in the nourishment of the body. The proteins are primarily tissue-builders and also help somewhat in force production. The fats are essentially heat-producers, though they too help in force production. The carbohydrates are chiefly important as force-producers, though they also produce heat and to a certain extent save protein oxidation. Fat is formed by all three but only in small amount by proteins. So no one food can form the whole diet but there must be variety. Carbohydrates and fats are not sufficient for life, some protein is necessary. Carbohydrates are more digestible than fats but have less potential energy. Gelatine saves waste of nitrogen, though it does not increase the supply. Water and salts are not nutritive but they aid the body processes, the water helping to dilute and dissolve substances for digestion.

The end-product of the consumption of protein is *urea*, which is eliminated by the kidneys. Just where it is formed is unknown, but many think in the liver. A trifling amount of urea is also eliminated in the sweat and in the breath as well as in the feces. Proteins increase nitrogenous metabolism and also the metabolism of other foods, but the amount of nitrogen eliminated is just equal to that taken in. Probably some comes from the tissues themselves and not from the food. The oxidation of carbohydrates and fats is measured by the amount of carbon excreted. At first as much is given off as is taken in, but after a while the carbohydrate is stored up as glycogen in the liver and the fats are stored as fat.

The amount of food needed varies with the person's size and occupation, less being needed for a child than for an adult and more for a hard-working man than for one who is doing less work. In general, 100 to 130 grams of protein, 40 to 80 grams of fat, 450 to 550 grams of carbohydrates, 30 grams of salts, and 28,000 grams of water is a fair amount.

Foods are cooked to make them more digestible and to develop their flavor, so that they will taste better. Cooking also kills germs and parasites that might be harmful. Meats should be cooked rapidly on the outside to coagulate the surface albumen and keep in the juices. The heat, besides coagulating the albumen, turns the tough parts to gelatine. In cereals the tough envelope of cellulose is broken up and in vegetables the tough fibrous parts are softened and made more digestible.

The Liver.—Below the diaphragm on the right and extending across above the stomach, resting in a way upon the transverse colon and the small intestine, is the liver, the largest gland in the body. It is dark reddish-brown in color and is larger in proportion in the child than in the adult. The upper surface is convex and lies in contact with the diaphragm, while the lower surface is concave to fit over the organs beneath. With a full breath it comes downward and forward, with the edge against the abdominal wall, and can be easily felt. Numerous strong ligaments, including the suspensory ligament from the diaphragm, hold it in place, and it is more firmly fixed than any other of the abdominal organs, probably on account of its large size. It is divided by fissures into five *lobes*, of which the most important are the right and left, the right one being the largest and containing the gall-bladder in one of its fissures.

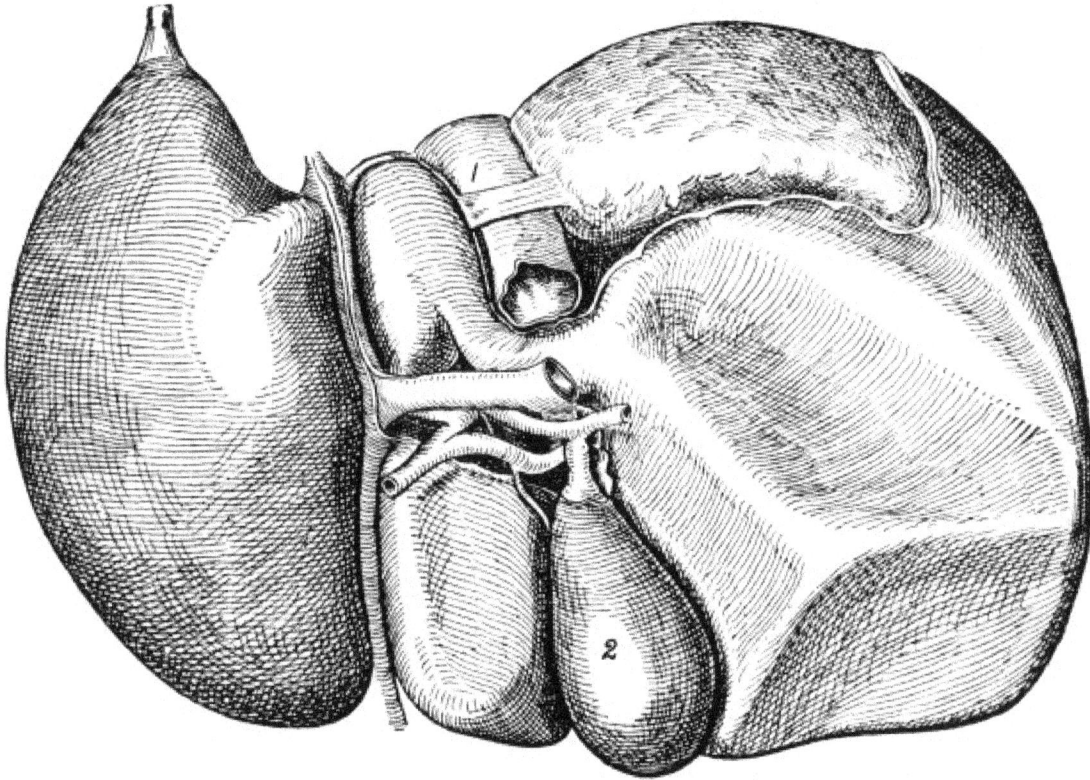

FIG. 56.—The liver, seen from below. 1, Inferior vena cava; 2, gall-bladder. (Morrow.)

The liver tissue contains a large number of *cells* collected into *lobules*, in the center of each of which is a blood-vessel, the *intralobular vein*, from which a network of capillaries extends to the edge of the lobule, there being a capillary on either side of each row of cells. Between the cells also are the *intercellular biliary passages*, roots of the bile ducts which exist in the connective tissue between the lobules and which join to form two main ducts, one from the right and the other from the left lobe. By the union of these two ducts the *hepatic duct* is formed, which, after a course of one or two inches, joins the cystic duct from the gall-bladder to form the *ductus communis* or *common bile duct*.

The liver has a double *blood supply*, the *hepatic artery* from the celiac axis bringing nourishment to the connective tissue and the walls of the blood-vessels, while the capillaries between the cells come from the *portal vein*, which, being formed by the junction of the superior and inferior mesenteric, the splenic and the gastric veins, contains the proteins and carbohydrates absorbed during digestion. After its passage through the liver this blood from the portal vein is collected once more into the hepatic veins, which convey it to the inferior vena cava. During its passage, however, various changes take place, for the liver plays an important part in the metabolic processes of the body.

The liver has two principal *functions*, the secreting of bile and the storing up of glycogen. The secretion of *bile*, which is a very important aid to digestion, is probably a reflex act, the presence of peptones in the portal blood after meals acting as a stimulant to the liver cells. For food at once increases the secretion of bile, which is poured from the cells into the small bile ducts and finally passes into the hepatic duct and so to the gall-bladder, where it is stored until needed. Although the flow from the liver is constant, the amount secreted reaches its maximum when the food gets down into the small intestine, that is, four or five hours after eating, there being a lull before that. Apart from the process of secretion, the manufacture of the bile pigments, *bilirubin* and *biliverdin*, which are made from the hemoglobin of the blood, seems to require some special action on the part of the liver cells.

The *glycogen*, which is manufactured and stored in the liver cells, is a clear hyaline substance, akin to starch and capable of being converted into sugar by the starch ferment. Probably there is some such ferment in the blood which converts the glycogen into sugar as soon as it passes from the liver into the blood, though what it is, is not known. Neither is it known just how glycogen is formed, but it is manufactured chiefly after a mixed meal in which carbohydrates predominate, proteins having little and fats no effect upon its formation. It is undoubtedly formed from the sugar in the portal blood and the process requires some work on the part of the liver cell itself. Probably there is always some sugar in the circulating blood which, as it is used up, must be made good. If there it not enough in the diet, the liver supplies the deficiency from its store of glycogen.

Glycogen is found also in the muscles, in the placenta as food for the fetus, in leucocytes, and to a slight extent in cartilage. In fact, it is the form in which carbohydrate material is supplied to the tissues as needed. Normally, much of the sugar is used up by the blood and its cells in metabolism, giving rise to heat and energy. In muscles glycogen is probably digested as lactic acid, as before action muscle is neutral or slightly alkaline and after action acid.

When the liver is deranged and allows the glycogen to pass out into the blood too freely, or when the glycogen is not held as such but turned to sugar and passed out in large quantities, *sugar in the urine* or *diabetes mellitus* results.

Besides its secreting function the liver has an *eliminative function* and plays an important part in purifying the blood, removing from it many poisonous and narcotic substances. It is thought by some, though it has not been proved, that *urea*, the end-product of protein metabolism, which is brought by the blood to the kidneys and there excreted, is formed in the liver. At any rate, urea is formed not only from the nitrogenous food eaten but from the metabolism of protein substances in the tissues, being purely a waste product, from which the nutritious substances have been absorbed. The amount thrown off is an accurate gauge of the amount of protein metabolism going on. The process of its manufacture is doubtless very complex.

Ptosis or *dropping* of the liver sometimes occurs and is due to the stretching of the ligaments. *Rupture* is common, generally as the result of a fall from a height, on account of its size and friability. The liver is also subject to many diseases. *Cirrhosis* occurs in people who drink a good deal and in its later stages is accompanied by *ascites*, an accumulation of fluid in the abdominal cavity. When there is a general accumulation of fluid throughout the body it is known as *anasarca*. *Syphilis* causes enlargement of the liver. *Abscesses* occur, perhaps oftener in the tropics than farther north, and may break into the lungs, stomach, or intestine.

The Gall-bladder.—The gall-bladder, which is simply a reservoir for the bile, is a pear-shaped organ three inches long and one inch broad. It lies in a fossa on the under side of the liver, with the large end or fundus touching the abdominal wall just below the ninth costal cartilage. Here it can be felt as a small mass in empyema of the gall-bladder. Normally it holds a little over one ounce, but with occlusion it may become stretched. Its duct is the *cystic duct*, which joins the hepatic duct in the common bile duct, but bile only passes up into the gall-bladder when the opening into the duodenum is closed, that is, between meals.

If one of the bile ducts is stopped up by a stone or cancer or for any other cause, the bile backs up in the liver, the pigments are absorbed into the circulation, and *jaundice* results. In this condition operation is dangerous, as the time of coagulation of the blood, normally five minutes or less, is much delayed. *Gall stones,* formed largely of bile pigments and cholesterin, sometimes collect in the gall-bladder, where they cause irritation and may give rise to *empyema* of the gall-bladder. The stones vary in size from a pea to a hen's egg and when small may be very numerous.

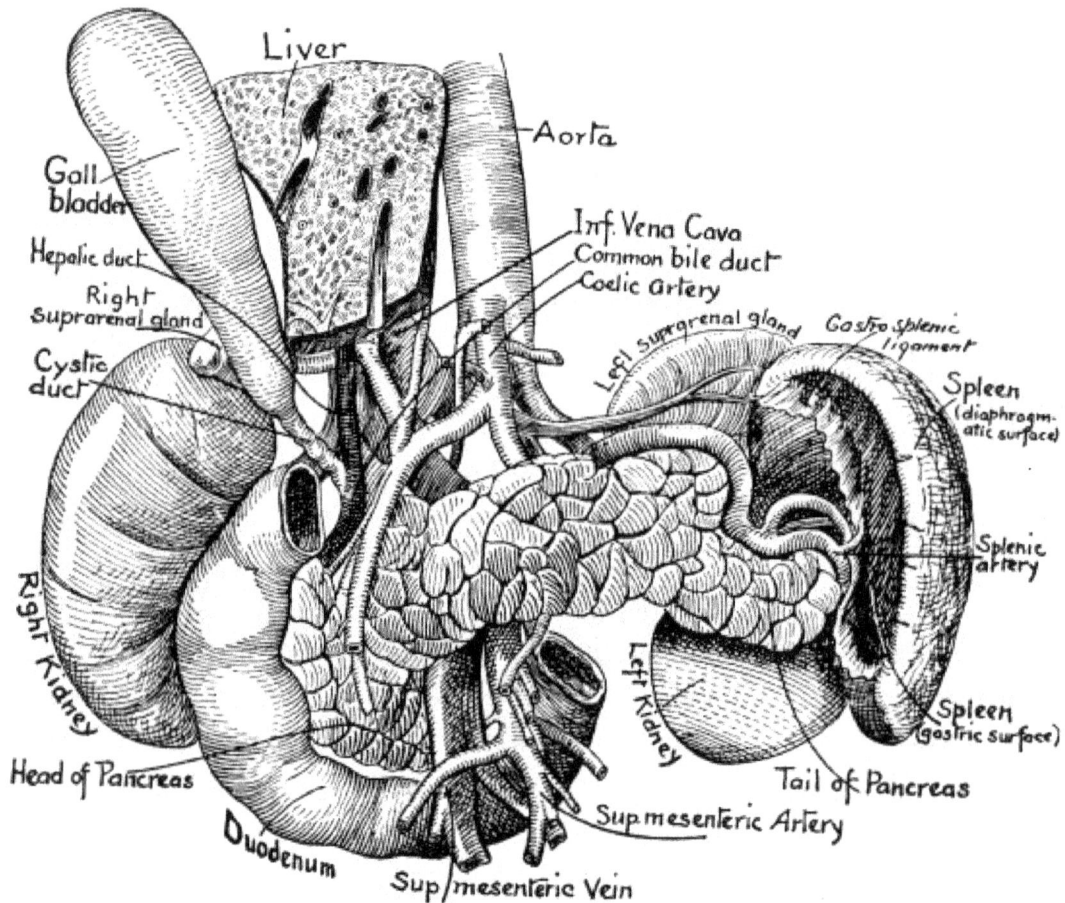

Fig. 57.—The pancreas, spleen, gall-bladder, etc., showing their relations. (After Sobotta.)

The Pancreas.—Another accessory organ of digestion is the pancreas, the *abdominal salivary gland,* as it is sometimes called on account of its close resemblance to the parotid gland. This is a grayish-white racemose gland, six and a half inches long by one and a half inches wide and one inch thick, lying behind the stomach on a level with the first and second lumbar vertebræ and shaped like a pistol with its handle toward the right. In an emaciated person it can be felt. The *pancreatic duct* runs the whole length of the gland from left to right and conveys the *pancreatic juice* from various little glands in the substance of the organ to the duodenum, into which it empties along with the common bile duct by a common orifice. The *arteries* are from the celiac axis and superior mesenteric, the *veins* belong to the portal system, and the *nerves* come from the solar plexus.

Surgically the pancreas is of no special importance, though *acute pancreatitis* does occasionally occur and is a very serious condition and one hard to diagnose.

The Spleen.—The largest and most important of the ductless glands is the spleen, an oblong, flattened organ lying deep in the left hypochondriac region between the stomach and diaphragm above the descending colon, and corresponding to the ninth, tenth, and eleventh ribs. It is soft, brittle, and very vascular. Its *artery* is a branch of the celiac axis and the *vein* belongs to the portal system. Its *nerves* are the pneumogastric and branches from the solar plexus. The *function* is not well understood but probably it is connected with or related to the vascular system in some way. Perhaps it manufactures blood corpuscles.

The spleen varies more in size than any other organ. Normally it cannot be felt, but in typhoid it usually can. It is generally atrophied in old age and hypertrophied in almost all acute infectious diseases, especially in typhoid fever and malaria. In leukemia it is often greatly enlarged. Sometimes in violent falls it is *ruptured* and there is considerable hemorrhage.

The Suprarenal Capsules.—The other ductless glands, the suprarenal capsules, yellowish triangular bodies, are situated just above and in front of the kidneys. Their function is important but not well understood. Death, accompanied by great muscular weakness, follows the removal of both, and when they are diseased, similar weakness is observed and the skin becomes bronzed. Injection of the extract of the suprarenals stimulates the muscular system. So probably they secrete into the blood minute quantities of a substance or substances beneficial to the body, especially to the muscular system.

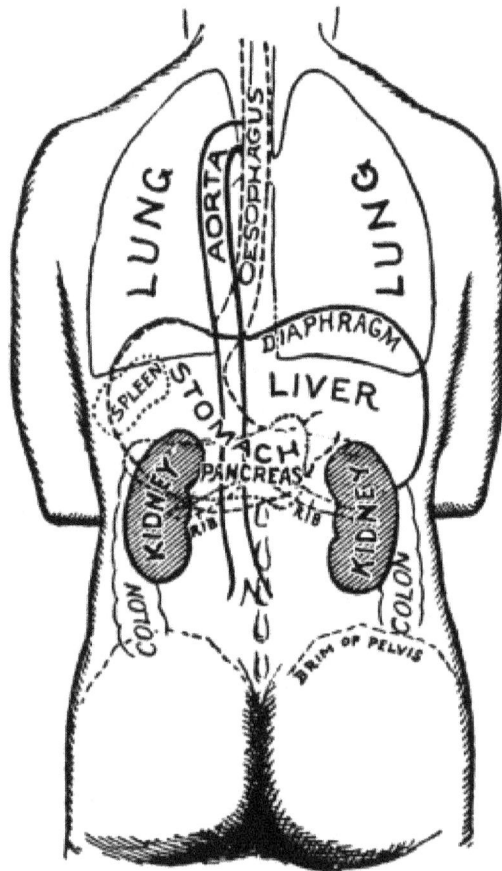

Fig. 58.—Diagram of the relation of kidney to viscera, spine, and surface points.
(American Text-Book of Surgery.)

The Kidneys.—The two kidneys lie on either side of the vertebræ at the back of the abdominal cavity and behind the peritoneum, between the last dorsal and the third lumbar vertebræ, their inner edge being about one inch from the spinous processes. They are bean-shaped, four inches long, two inches wide, and one inch thick, and are embedded in a mass of fat and loose areolar tissue. They can be felt only when misplaced or when enlarged, as by tuberculosis or malignant disease.

The whole kidney is enveloped in a fibrous *capsule* which normally may be peeled off but which in some diseases becomes adherent. On the internal border is a *fissure* or *hilum*, through which pass the blood-vessels and the ureter. Upon entering, the ureter dilates into a sac, the *pelvis of the kidney*, into which project the *Malpighian pyramids* of the *medullary substance*, a substance made up of the straight uriniferous tubules and blood-vessels. Outside the medullary substance and just under the capsule is the *cortex*, containing the *Malpighian bodies*, blood-vessels, and the *convoluted tubules* or *loops of Henle*. Each Malpighian body contains within a capsule a plexus of capillaries, the *glomerulus*, with an afferent arteriole and an efferent vein. The *renal artery* is a branch of the aorta and the *nerves* are from the solar plexus.

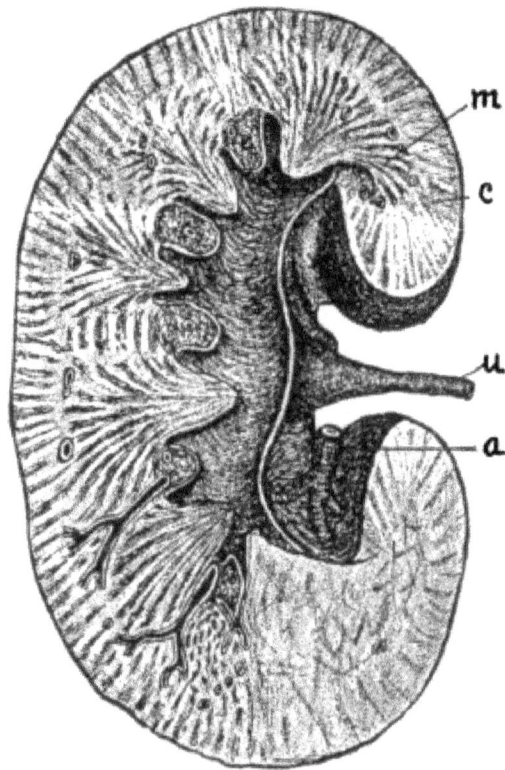

FIG. 59.—A longitudinal section
of the kidney. (Leroy.)
a, Renal artery; *c*, cortex;
m, medulla; *u*, ureter.

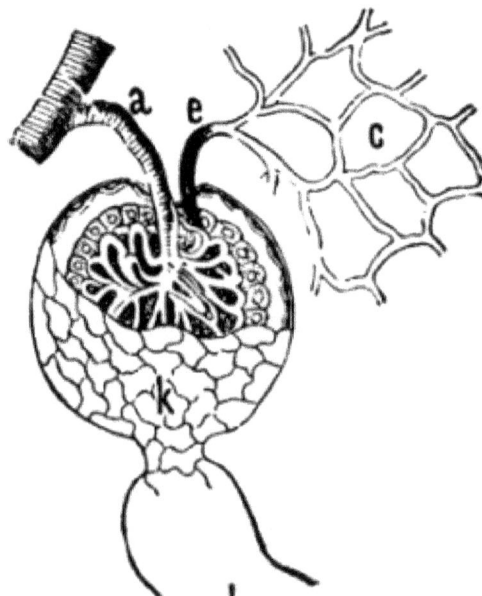

FIG. 60.—A Malpighian body
or corpuscle. (Leidy.)
a, Afferent artery;
e, efferent vessel;
c, capillaries;
k, commencement of
uriniferous tubule;
h, uriniferous tubule.

The Urine.—As the blood passes through the glomeruli, the urine is filtered off as it were, probably by a process of transudation rather than simple filtration. The cells lining the tubules also play an important part in its formation, not by secreting new substances but by taking up those brought by the blood and discharging them into the convoluted tubules, from which the urine passes through the straight tubules of the medulla to the pelvis, to be carried thence by the ureter. The process of the formation of the urine, therefore, is not purely a process of secretion but requires some action on the part of the kidney, though no new substances are secreted in the kidney.

The passage of the urine down through the ureters is assisted by a kind of peristaltic action in the walls of the ureters and it is expelled from the body by the act of *micturition*, which is mostly voluntary, though a certain amount of nervous mechanism controls it. The seat of this nervous mechanism is in the lumbar enlargement of the spinal cord. In some nervous conditions, especially where there is injury to the spinal cord, there is *involuntary micturition*.

The urine is a watery solution containing many waste products, especially urea. It is generally amber in color, varying in shade with circumstances, with an aromatic, characteristic odor when fresh. It is acid in reaction and has a specific gravity of about 1020, though this too varies with circumstances. Besides water, which is its chief constituent, it contains urea, uric acid, organic acids, urates, inorganic salts, including sodium chloride and phosphates of calcium and magnesium, a certain amount of ammonia, and certain pigments. Its acidity is due to acid sodium phosphate in solution but varies with the food, and in disease the urine may become alkaline when passed. After standing a few hours in a warm place it decomposes and becomes alkaline.

The quantity, which is normally three pints or fifteen hundred cubic centimeters in twenty-four hours, varies with the amount of fluid drunk, the amount of perspiration, etc. The amount secreted depends chiefly, however, upon the flow of the blood through the kidneys; the greater the flow of blood, the larger the amount of urine formed; and the blood flow is determined by blood pressure and by vasomotor action. Secretion also seems to be increased by the presence of urea, which apparently serves as a stimulant to the kidney cells.

The excretion of waste materials takes place by three main channels, the lungs, skin, and kidneys, and the materials are of four kinds, *urea*, *carbon dioxide*, *salts*, and *water*. The lungs carry off carbon dioxide and water chiefly, the skin these and inorganic salts, while the kidneys eliminate practically all the urea as well as inorganic salts and water. When the kidneys are not working the skin carries off much urea. In fact, a close relationship exists between the kidneys and the skin in the matter of excretion. Thus, with increased perspiration in warm weather comes decreased urine, while in cold weather the blood is sent in and the urine increased in amount.

To incite action of the kidneys drugs known as *diuretics* may be used. These act in two ways, by stimulating the kidney cells directly and by acting on the general circulation or nervous system. Any emotional or nervous excitement increases the flow of urine.

There are certain abnormal constituents of urine, of which the two most important are *albumen* and *sugar*. The former is found only when there is some disturbance of the kidneys, ureters, or bladder, and its presence usually denotes some change in the cells lining the urinary tract. It may occur in congestion of the kidney as well as in disease. Sugar is found only in *diabetes*, the amount varying with the severity of the disease. In *jaundice* certain bile pigments are present in the urine, giving it a dark brown color and to the foam a greenish-yellow color. Even normal urine has some sediment upon standing, consisting of cells from the urinary tract and mucus. In very acid urine after standing a heavy sediment, whitish or pinkish, *i.e.*, brick dust, in color, is thrown down. It does not necessarily denote disease, but shows the urine is acid and concentrated. In *alkaline urine* there is a sediment due to phosphates.

Rupture of the kidney occurs but is not so serious as rupture of the liver or spleen because the kidney is situated outside of the peritoneum. It necessitates the removal of the kidney, however, and when for any reason one kidney is removed the other increases in size and does double work to compensate for the loss. Removal of both kidneys means death. Sometimes the kidney becomes loose and moves about, a condition known as *floating kidney*. *Perinephritic abscess* is abscess in the loose fatty tissue about the kidney.

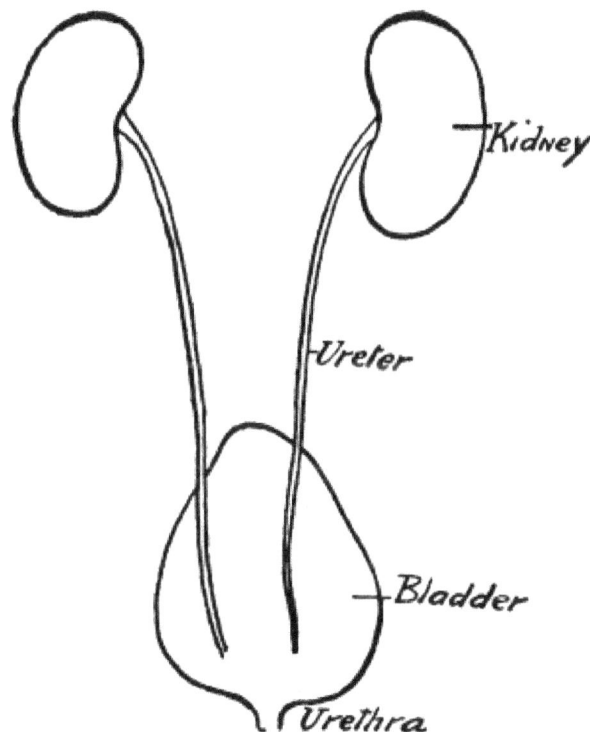

FIG. 61.—The urinary organs
viewed from behind.

The Ureters, one for each kidney, are tubes the size of a goose quill and about fourteen inches long, extending from the hilum of the kidney to the base of the bladder. They have three *coats*, an internal mucous, a muscular, and an external fibrous coat, this last being continuous with the cortex of the kidney and the fibrous tissue of the bladder. In the female the ureters may be felt through the wall of the vagina as they come into the bladder. In tubercular disease of one kidney the ureter becomes inflamed and enlarged and through the vagina feels almost like a lead pencil, a sure diagnostic sign.

The Bladder and Urethra.—In their course to the bladder the ureters pass from the abdominal into the pelvic cavity, but before describing the pelvis itself it will be well to complete the account of the urinary organs by considering the bladder and urethra. The *bladder* is the reservoir for the urine and has muscular walls lined with mucous membrane. A peritoneal coat covers the upper surface and is reflected to the walls of the abdomen and pelvis. It is situated back of the os pubis, the front bone of the pelvis, with its base or fundus directed downward and backward. Normally it is in the pelvis, but when much distended it mounds up into the abdominal cavity, where it can be felt in front as a tumor. It rests on the rectum in the male and on the cervix in the female and is held in place by numerous ligaments. When empty it may be Y-shaped, but it becomes oval when distended. Its capacity is about one pint.

The lower abdominal wall and the anterior wall of the bladder may be wanting congenitally. In paralysis of the sphincter at the neck of the bladder distention results. *Stones* may be found in the bladder.

From the neck of the bladder the urine passes out of the body through the *urethra*. This in the male passes down through the penis and is about ten inches long. Except when urine is passing it is a transverse slit with the upper and under surfaces in contact, while at the end of the penis the slit of the meatus urinarius is vertical. When the penis is flaccid, the urethra describes a sharp curve before its entrance into the bladder, but it becomes approximately straight when the penis is raised at right angles to the body—an important point to remember in catheterization.

In the female the urethra is straight and much shorter, being only about one and a half inches long. The *meatus urinarius* is in the anterior vaginal wall about one inch behind the clitoris.

Sometimes the urethra is *ruptured* in a fall. *Stricture* of the urethra occurs sometimes after gonorrhoea, owing to the formation of scar tissue following ulcer.

CHAPTER XI.
THE PELVIS AND THE
GENITAL ORGANS.

The Pelvis.—Before taking up the pelvic organs, the pelvis itself should be described. The name pelvis has been given to the bony ring which is interposed between the spine and the femurs on account of its resemblance to a basin. At the back of this basin or pelvis are the *sacrum* and *coccyx*, already described in connection with the back, and at the sides and meeting in the median line in front are the two *ossa innominata* or *nameless bones*, so called on account of their peculiar and indescribable shape. At birth each os innominatum is made up of three bones, the *ilium*, *ischium*, and *pubes*, but about the age of puberty the three become welded into one. At their point of junction is the cavity of the acetabulum for articulation with the head of the femur or thigh bone.

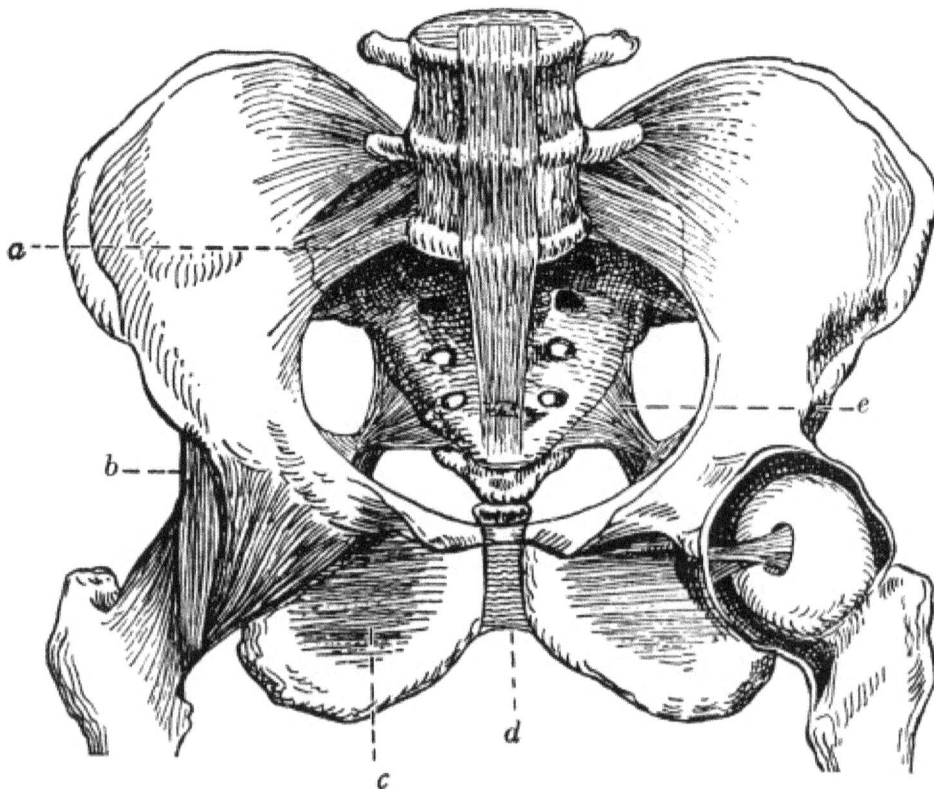

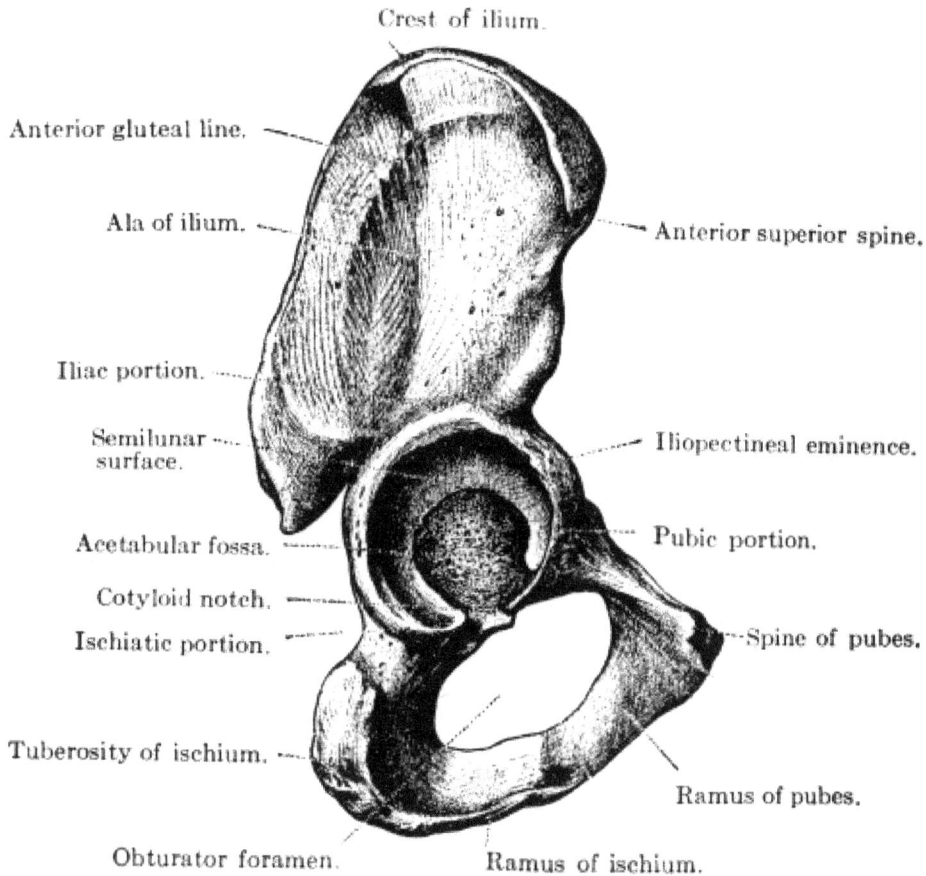

Crest of ilium.

Anterior gluteal line.

Ala of ilium.

Anterior superior spine.

Iliac portion.

Semilunar surface.

Iliopectineal eminence.

Acetabular fossa.

Pubic portion.

Cotyloid notch.

Ischiatic portion.

Spine of pubes.

Tuberosity of ischium.

Ramus of pubes.

Obturator foramen.

Ramus of ischium.

FIG. 63.—The right innominate bone.
(After Toldt.)

The upper, expanded portion of the os innominatum is the *ilium*, whose upper border is known as the *crest* and which has two *spinous processes* front and back, a superior and an inferior, the superior spine being in each case the larger. These spines, especially the anterior superior spines, and the crest give attachment to many muscles, and to the outer surface of the bone the gluteal muscles are attached. The anterior superior spine is also important in making measurements to ascertain whether both legs are of equal length.

Below the ilium posteriorly is the *body of the ischium*, which has on its lower edge a tuberosity, the prominent bone on which one sits. Near the upper edge is the *spine* of the *ischium*, between which and the posterior inferior spine of the ilium is the greater *sacro-sciatic notch* for the passage of vessels and nerves, including the sciatic nerves. From the tuberosity the ramus extends forward below the *obturator foramen,* a large opening between the ischium and the pubes, also for the passage of vessels and nerves, to meet the pubes, the last and smallest of the three bones which go to make up the os innominatum.

The anterior surface of each *pubes* presents a crest, ending externally in a spine, and the two pubic bones join in front in the *symphysis pubis.* The bone gets its name from the growth of pubic hairs over this region at puberty.

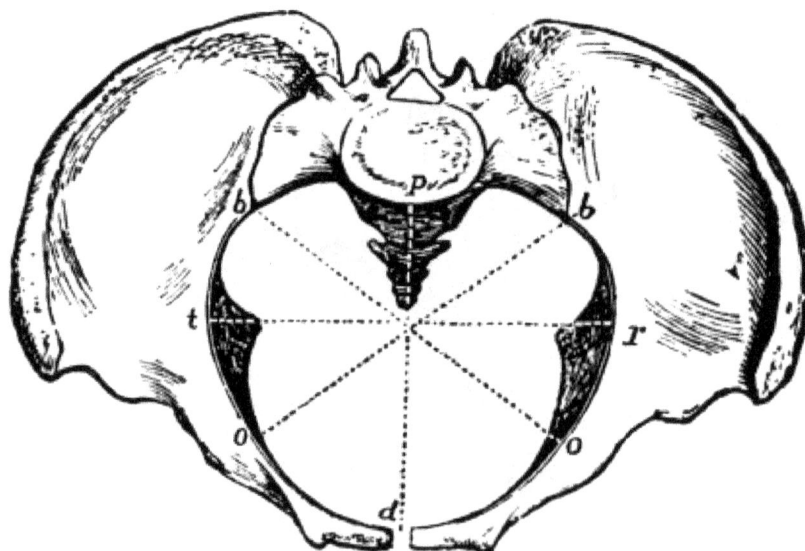

Fig. 64.—Diameters of the pelvis: *d,* antero-posterior; *o b,* oblique; *t r,* transverse. (de Nancrede.)

Anteriorly the ossa innominata support the external organs of generation, while within are the internal organs of generation. On the inner surface of the ilium, slightly above the level of the acetabulum, is the *ileo-pectineal line,* above which lie the *iliac fossæ.* A plane drawn through the prominence of the sacrum, the ileo-pectineal lines, and the upper margin of the symphysis pubis serves to divide the upper or false pelvis from the lower or true pelvis. The *false pelvis,* which is the larger, serves to support the intestines and to take part of the weight from the abdominal walls, while the *true pelvis,* being more surrounded by bone and so capable of affording more protection, guards the internal organs of generation. The lower circumference of the pelvis is known as the *outlet.* In the female the bones are lighter, the sacrum less curved, and the diameters greater than in the male.

On the whole, the pelvic bones are well covered in with muscles. The anterior superior spine, however, is easily felt in front and the whole crest can be felt back to the posterior superior spine. The tuberosity of the ischium also can be felt, especially when the thigh is flexed, for it is largely uncovered of muscles. The spine of the os pubis can always be felt, on a level with the great trochanter, and the relation of its position to that of a hernia shows whether the rupture is above or below Poupart's ligament, that is, whether it is inguinal or femoral.

Occasionally there is lack of development of the pubic bones for two or three inches and the bladder is exposed. *Fracture* of the pelvis may occur, perhaps with injury to the viscera. The acetabulum may be fractured or the sacrum broken, with injury to the sacral plexus of nerves, causing paralysis of the lower extremities and of the sphincters, with resultant involuntary passage of urine and feces, and in childbirth the coccyx is often broken. In *rickets* there may be great deformity of the pelvis, causing trouble in childbirth later in life. *Osteomalacia* is a disease of adults, in which the bones are soft and the weight pushes the promontory of the sacrum forward and approximates the sides of the pelvis.

The Male Generative Organs.—The male generative organs consist of the prostate gland, testes, and penis.

The *prostate gland* is shaped like a small horse-chestnut and is composed of numerous glands from which come a dozen or more excretory ducts. It surrounds the neck of the bladder and the beginning of the urethra and is next to the rectum, through which an examination may be made to determine its size. For it often enlarges in elderly men, the frequent passage of urine in small amounts being a symptom of enlarged prostate.

The *procreating glands*, which secrete the spermatozoa or semen, are two in number, the *testes* or *testicles*, and are homologous to the ovaries in the female. They are ovoid in form and are suspended by the *spermatic cords* in a sac, the *scrotum*, back of the penis. During early fetal life they are in the back of the abdomen near the kidneys, but before birth they descend along the inguinal canals into the scrotum. The excretory duct of the testis is called the *vas deferens*. It passes up by the spermatic cord through the inguinal canal into the pelvis to the base of the bladder and at the base of the prostate joins the duct of the vesicula seminalis to form the *ejaculatory duct.* The two *vesiculæ seminales* are small receptacles for the semen lying in contact with the base of the bladder and secrete a fluid with which they dilute the semen. The ejaculatory duct terminates near the prostate in the urethra by a slit-like orifice on each side, the spermatozoa being finally excreted through the urethra.

The *penis* is the external organ of generation in the male and is attached to the pubes and the anterior part of the ischium. It is composed of erectile tissue and encloses the urethra, the *meatus urinarius* appearing at its end as a vertical slit. Toward the end the skin of the penis is loose and is prolonged forward in what is known as the *prepuce* or *foreskin*. It is this that is clipped away in circumcision.

The Female Generative Organs.—The female generative organs include the ovaries, Fallopian tubes, uterus, vagina, and the external genitalia or vulva.

The <u>ovaries</u>, which are homologous to the testes in the male, are two flattened oval bodies, grayish pink in color, suspended from the lateral or broad ligaments which fasten the uterus to the walls of the pelvis. They are one and a quarter inches long, three-quarters of an inch wide, and half an inch thick and are attached at the upper end to one of the fimbriæ of the <u>Fallopian tubes</u>. They consist of numerous *Graafian follicles* embedded in a fibrous stroma, each follicle containing an *ovum* about $1/125$ inch in diameter and just visible to the naked eye. When a follicle ruptures and discharges an ovum, an irregular yellow spot, the *corpus luteum*, appears at the point of rupture. After ordinary menstruation it is known as the *false corpus luteum* and after conception as the *true* one, this one being larger and lasting longer.

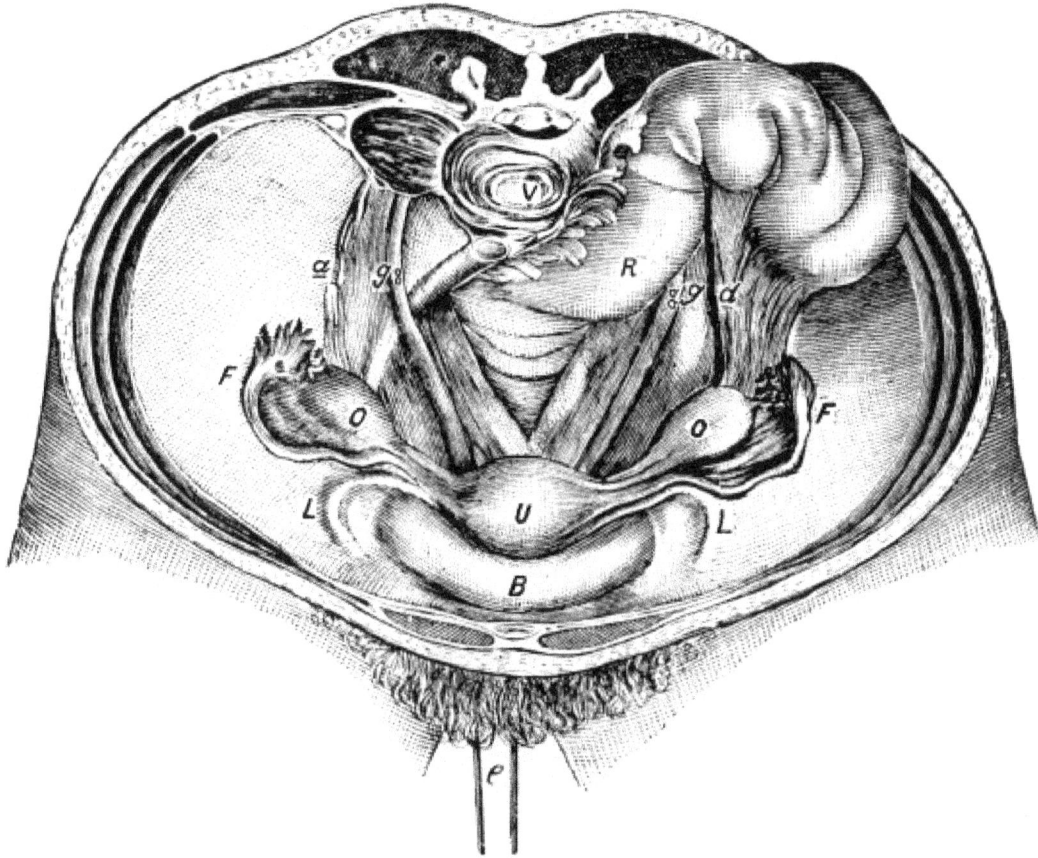

FIG. 65.—View of the pelvis and its organs. (Savage.) *B*, Bladder;
U, uterus (drawn down by loop *e*); *F*, Fallopian tubes; *O*, ovaries; *L*,
round ligaments; *g*, ureter; *a*, ovarian vessels, often prominent under
their peritoneal covering; *R*, rectum; *V*, vertebra.

The *Fallopian tubes* are the oviducts and convey the ova from the
ovaries to the uterus. They are four inches long and lie between the layers
of the broad ligaments, opening into the uterus by an orifice the size of a
bristle, while the end next to the ovary spreads out trumpet-like and is
edged with *fimbriæ* as with a fringe, the fimbriated extremity. There are
three *coats*: a serous coat which is continuous with the peritoneum, a
muscular coat, and, within, a mucous coat covered with cilia, continuous
with the mucous membrane of the uterus. One fimbria is attached to the
ovary and as the ovum is given off it finds its way into the tube and thence
to the uterus.

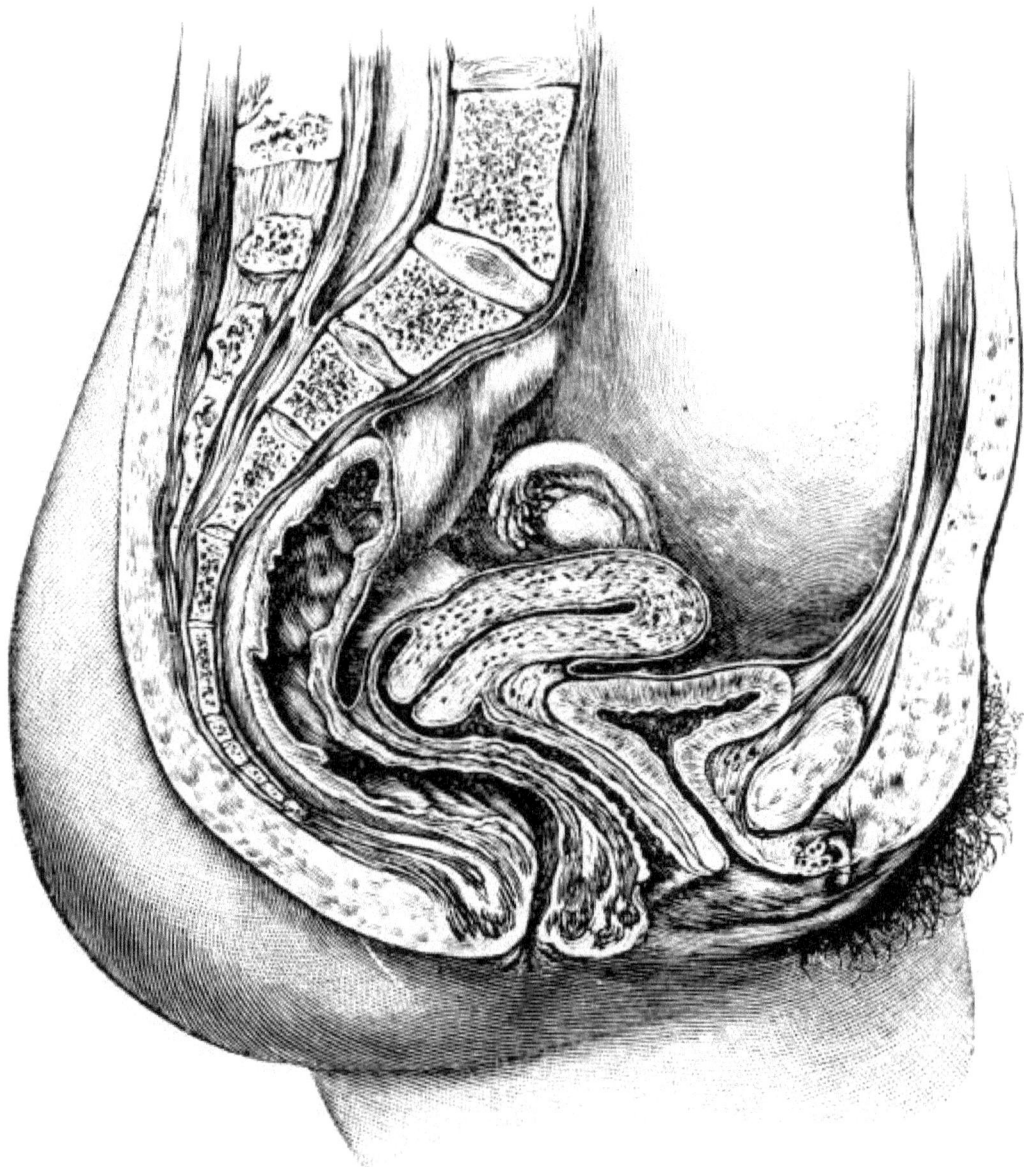

FIG. 66.—Sagittal section of the female pelvis.
(Dickinson.)

The <u>uterus</u> is a pear-shaped organ, about three inches long, two inches broad above, and one inch thick, situated in the pelvic cavity between the rectum and the bladder. The wide part or *fundus* is above and the narrow neck or *cervix* below, lying partly within the vagina. The whole is held in place by ligaments. These include the *broad ligaments*, which extend from the sides of the uterus to the lateral walls of the pelvis, and the *round ligaments*, two muscular cords, about four inches long, which pass out through the abdominal ring into the inguinal canal and so to the mons veneris and labia, thus corresponding to the spermatic cords in the male.

The cavity of the body of the uterus is small and flattened and opens into the cervix by the *internal os uteri*, the *external os* being at the opening of the cervix into the vagina. There are three *coats*: a serous coat derived from the peritoneum, a muscular coat of unstriped fibers which forms the bulk of the whole organ, and a mucous coat covered with ciliated epithelium.

The uterus is always enlarged during menstruation and is enormously enlarged in pregnancy. It receives the fecundated ovum, retains and supports it during the development of the fetus, and is the chief agent of expulsion. In *tubal* or *extra-uterine pregnancy* the ovum becomes attached in the tube instead of in the uterus, and develops there, rupturing the tube and causing serious hemorrhage.

The passage from the cervix out of the body is the *vagina*, a membranous canal, curved upward and backward to conform to the axis of the pelvis, and attached above to the cervix. Ordinarily the sides are in contact.

The *arteries* of the internal organs of generation are the uterine from the internal iliac and the ovarian from the aorta in the female, the pudic branches of the internal iliac and the spermatic from the aorta in the male. The *nerves* are largely from the sympathetic system.

Abscess formation occurs frequently in the tubes and *gonorrheal infection* may spread up the vagina and through the uterus to the tubes, and even to the abdominal cavity itself. The tubes may also be *tubercular*.

Salpingectomy or removal of the tubes is the commonest operation after that for appendicitis. *Cancer* of the uterus may necessitate *panhysterectomy* or removal of the uterus and all its appendages.

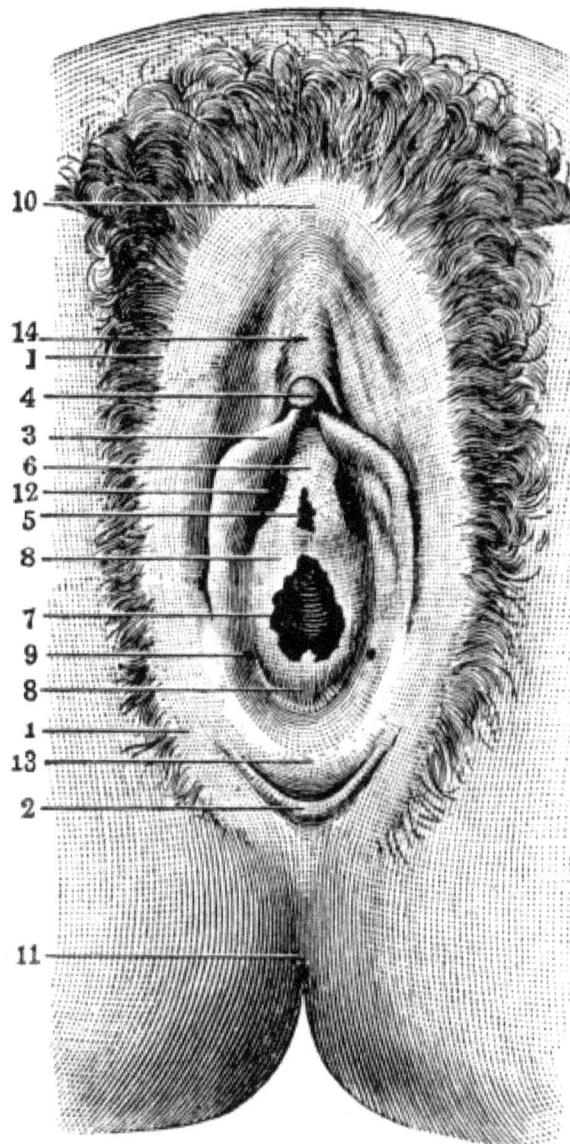

FIG. 67.—Virginal vulva. (Modified from Tarnier.) 1, labia majora; 2, fourchette; 3, labia minora; 4, glans clitoridis; 5, meatus urinarius; 6, vestibule; 7, entrance to the vagina; 8, hymen; 9, orifice of Bartholin's gland; 10, anterior commissure of labia majora; 11, anus; 12, blind recess; 13, fossa navicularis; 14, body of clitoris.

The *external genitalia* in the female, as in the male, are situated over the pubic arch. They are known as the <u>vulva</u> and include the mons Veneris, the labia majora and minora, the vaginal orifice, the clitoris, and the meatus urinarius.

The *mons Veneris* is a rounded eminence composed of fatty tissue, which surmounts the pubic bones and is covered with hair at puberty. From it two prominent longitudinal folds of skin, covered with hair on the outside, the _labia majora_, extend backward, forming the lateral boundaries of the vulva. Within these labia again are two thin cutaneous folds, the _labia minora_ or _nymphæ_, which run back from the clitoris for about one and a half inches and enclose the _vaginal orifice_. The _clitoris_ corresponds to the penis and is just above the upper part of the labia minora. Between it and the vagina is the _meatus urinarius_. The orifice of the vagina is partly closed in the virgin by the _hymen_, a thin fold of mucous membrane, which occasionally closes it completely, _imperforate hymen_. The _fourchette_ is a small transverse fold of skin at the junction of the labia minora posteriorly. Between the vagina and the rectum is the _perineal body_, a somewhat triangular structure made up of many small muscles. Its surface is known as the _perineum_. It is frequently torn wholly or in part during childbirth and has to be sewed up.

CHAPTER XII.
THE UPPER EXTREMITIES.

The upper extremities include the shoulders, arms, forearms, wrists, and hands and contain each thirty-two bones. The bones of the two shoulders taken together are called the *shoulder girdle* and consist of the two clavicles or collar bones and the two scapulæ or shoulder blades, which together make an almost complete girdle of the shoulders.

The clavicle is a long slender bone extending almost horizontally from the sternum to the scapula and can be felt for its whole length in the living. For the inner two-thirds it is convex anteriorly, for the outer third concave. In woman it is generally less curved, smoother, and more slender than in man, and as bone is rough when the muscles attached are powerful, the right clavicle, being used more, is generally rougher and thicker than the left. Among the muscles attached are the large neck muscle, the *sterno-cleido-mastoid*, whose tendons form the *presternal notch*, the *trapezius*, the *pectoralis major*, and the *deltoid*.

Being slender and superficial the clavicle is most frequently *broken* of any bone in the body, generally by indirect violence, as by falling with the hand out, though old people in such a case are apt to get *Colles' fracture* at the wrist. The bone generally gives way at the juncture of the outer and middle thirds, with displacement of the parts inward, so that the fracture is seldom compound. Since, however, the main vessels of the upper arm, with their nerves, lie beneath the clavicle, there is danger of their being punctured. Such serious injury is guarded against by the presence of the *subclavius muscle*. The clavicle is occasionally removed for *sarcoma*.

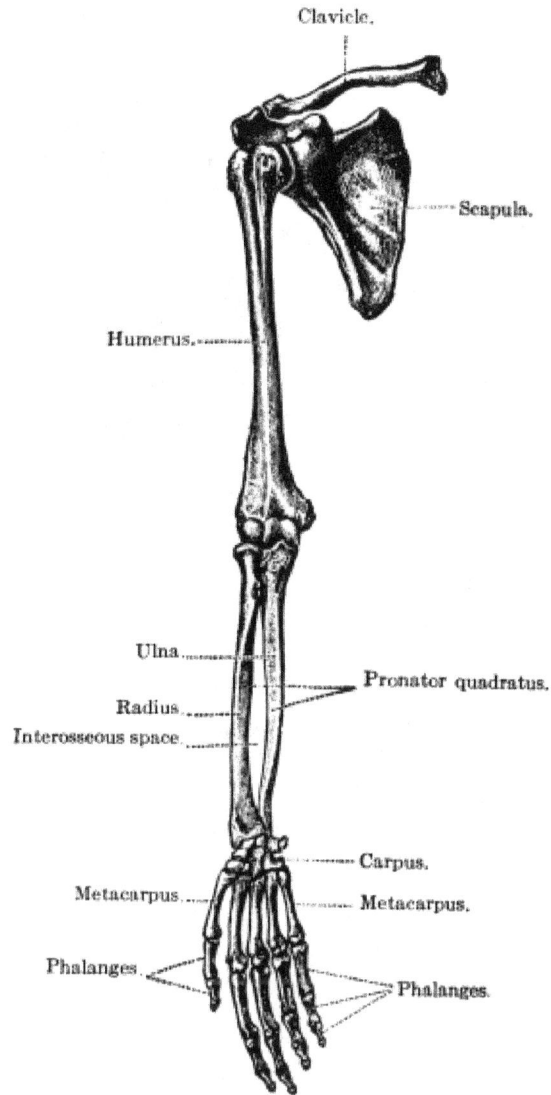

FIG. 68.—Bones of the upper extremity.
(Toldt.)

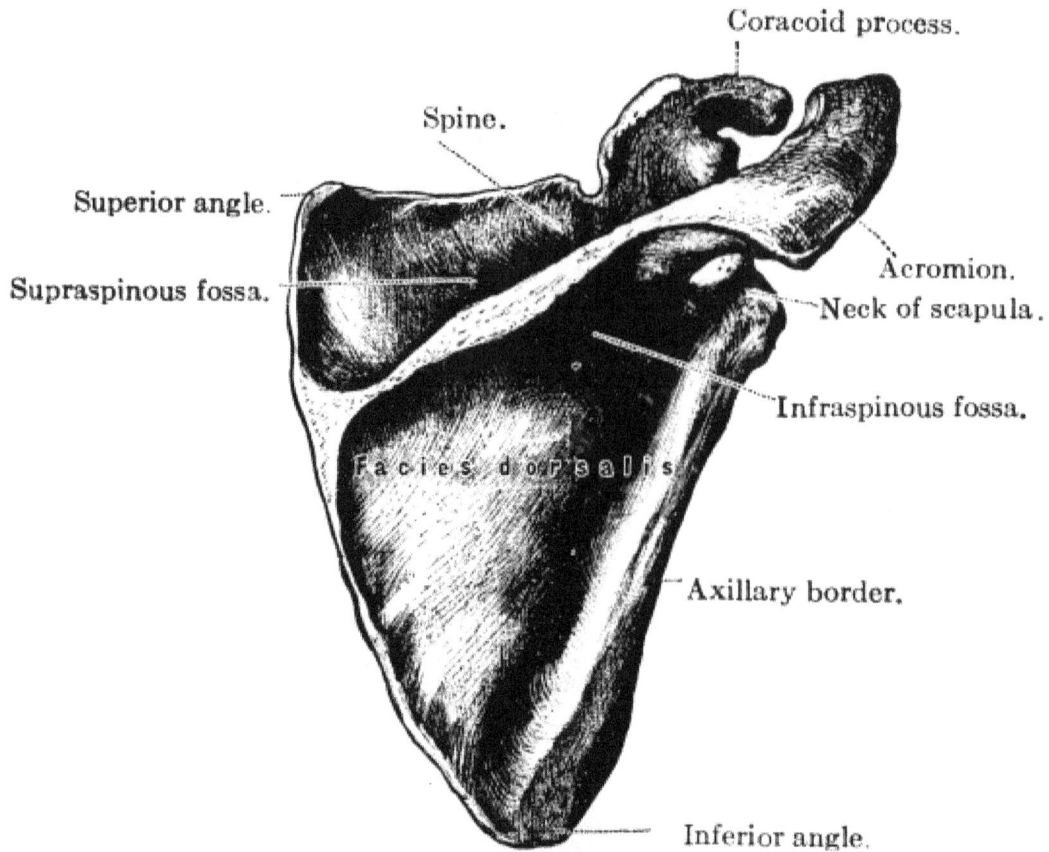

Coracoid process.

Spine.

Superior angle.

Supraspinous fossa.

Acromion.

Neck of scapula.

Infraspinous fossa.

Facies dorsalis

Axillary border.

Inferior angle.

FIG. 69.—Left scapula, posterior surface
(after Toldt).

The scapula or **shoulder blade**, so called from its shape, is a large, flat, triangular bone with a prominent ridge, the *spine*, crossing its dorsum or posterior surface near its upper edge. It extends from the second to the seventh rib, with its posterior margin parallel to and about one inch from the dorsal vertebræ. The *head*, in which is situated the *glenoid cavity* for articulation with the humerus or upper arm bone, is surrounded by a slight constriction, the *neck*. Above it projects the *coracoid process*, so called from its fancied resemblance to a crow's beak. This can usually be felt about one inch from the juncture of the outer and middle thirds of the clavicle and from it arise the short head of the biceps and the coracobrachialis muscle. The *acromion process* at the end of the spine extends out beyond the glenoid cavity posteriorly and affords attachment to the deltoid and trapezius muscles. It forms the summit of the shoulder. Numerous other muscles are attached to the surface of the scapula, the only parts which are truly subcutaneous being the whole length of the spine and the acromion process, though the lower angle and the coracoid process can generally be felt. The muscles bulge so much that the spine in the living appears as a slight depression extending back almost to the vertebræ. The large number of the muscles on the shoulder and arm is due to the great flexibility and strength required for the various uses to which the arms are put.

Shoulder Muscles.—The most important shoulder muscle is the _deltoid_, a large triangular muscle, which surrounds and protects the shoulder-joint and gives the shoulder its rounded form. It rises from the outer third of the _clavicle_, from the acromion process, and from the whole length of the spine of the scapula, and is inserted by a tendon into a rough prominence on the middle of the outer side of the humerus. It serves to raise the arm and to draw it somewhat forward or back, according as the anterior or posterior fibers are used. The _pectoralis major_ rises from the inner half of the clavicle, the front of the sternum, and the cartilages of the true ribs and its fibers converge to form a fan-shaped muscle, which is inserted by a flat tendon into the edge of the bicipital groove on the humerus. It draws the arm forward and inward and helps considerably in forced inspiration. The _serratus magnus_ rises from the outer surface and upper border of the eight upper ribs and from an aponeurosis covering the upper intercostal spaces, and is inserted along the whole length of the posterior border of the scapula. It carries the scapula forward and is used in pushing.

The scapula is seldom _broken_ because it is quite movable and is covered with large muscles and because it lies on the chest, which serves as an elastic cushion. The acromion process is the part most frequently broken and occasionally the neck is fractured. _Tumors_ occur and may necessitate the amputation of the whole upper extremity.

The Humerus.—The bone of the upper arm, the humerus, is the largest bone in the upper extremity and articulates with the scapula above and with the ulna and radius below. At its upper end are the _head_ and the _anatomical neck_, with the _greater tuberosity_ external to and the _lesser tuberosity_ in front of them. The constriction of the surgical neck is below the tuberosities, and extending from between them downward and inward along the upper third of the bone is the bicipital groove for the long head of the biceps. Though round above, below the _shaft_ becomes flattened from before backward and curves slightly forward, terminating in the internal and external _condyles_, from the former of which the flexors and the round pronator arise and from the latter the extensors and supinators. From the external condyle also there projects in front the *radial head* or *capitellum* for articulation with the radius. Internally to the capitellum in front and in a corresponding position on the back of the bone are the trochlear surfaces for articulation with the ulna, there being a depression in front called the *coronoid fossa* for the reception of the *coronoid process* of the ulna in flexion of the forearm, and another depression behind, the *olecranon fossa*, to receive the tip of the olecranon process during extension. On the lower half of the humerus at the back is the spiral groove for the *musculo-spiral nerve* and the *superior profunda artery*, while the *ulnar nerve* runs in a groove back of the internal condyle.

The humerus is almost completely covered with muscles, the only part that is subcutaneous being a small portion of the external and internal condyles. The head can be felt under the muscles and the greater tuberosity forms the point of the shoulder. When the arm is at the side, the biceps appears at the front and inner side and the brachialis anticus on either side below, while on the back of the arm, with its largest swelling above, is the triceps.

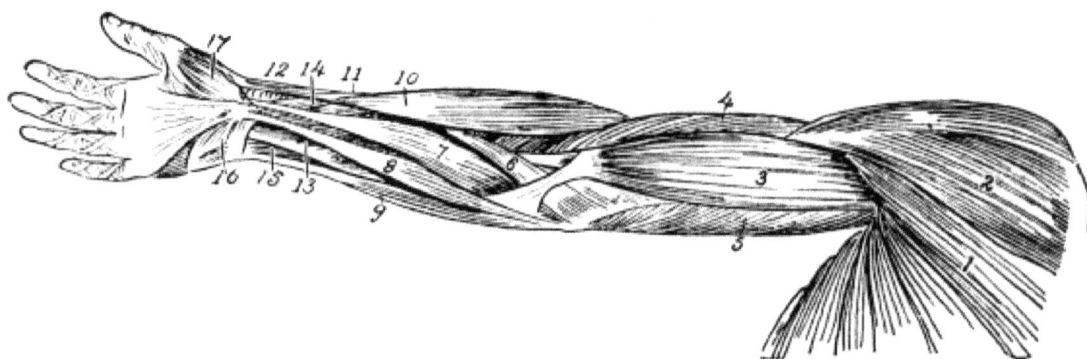

FIG. 70.

FIG. 70.—Superficial muscles of shoulder and arm (from before): 1, Pectoralis major; 2, deltoid; 3, biceps brachii; 4, brachialis anticus; 5, triceps; 6, pronator radii teres; 7, flexor carpi radialis; 8, palmaris longus; 9, flexor carpi ulnaris; 10, supinator longus; 11, extensor ossis metacarpi pollicis; 12, extensor brevis pollicis; 13, flexor sublimis digitorum; 14, flexor longus pollicis; 15, flexor profundus digitorum; 16, palmaris brevis; 17, abductor pollicis. (Dorland's Dictionary.)

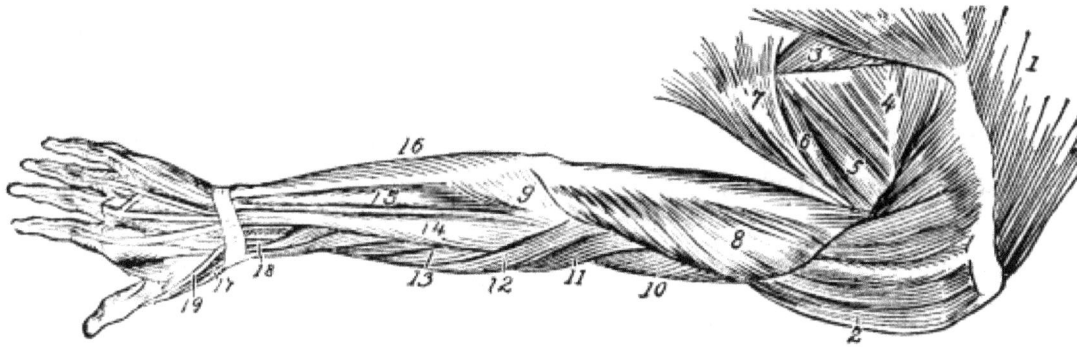

FIG. 71.

FIG. 71.—Superficial muscles of shoulder and arm (from behind): 1, Trapezius; 2, deltoid; 3, rhomboideus major; 4, infraspinatus; 5, teres minor; 6, teres major; 7, latissimus dorsi; 8, triceps; 9, anconeus; 10, brachialis anticus; 11, supinator longus; 12, extensor carpi radialis longior; 13, extensor carpi radialis brevior; 14, extensor communis digitorum; 15, extensor carpi ulnaris; 16, flexor carpi ulnaris; 17, extensor ossis metacarpi pollicis; 18, extensor brevis pollicis; 19, tendon of extensor longus pollicis. (Dorland's Dictionary.)

Upper Arm Muscles.—The *biceps* is the most important arm muscle. It rises by a short head from the coracoid process of the scapula and by a long head from a tubercle on the upper margin of the glenoid cavity, the tendon arching over the head of the humerus and descending in the bicipital groove. It is inserted into the back of the tuberosity of the radius and by a broad aponeurosis into the fascia of the forearm. It flexes and supinates the forearm and renders the fascia tense. Its inner border forms a guide in tying the *brachial artery*, as this artery runs along its inner side.

The *brachialis anticus* rises from the lower half of the outer and inner surfaces of the humerus and is inserted into the coronoid process of the ulna, thus covering and projecting the elbow-joint anteriorly. It is a flexor of the forearm.

Another smaller muscle on the anterior arm, which also aids in flexion, is the *coraco-brachialis,* which extends from the coracoid process of the scapula to the middle of the inner surface of the humerus.

Extending the entire length of the posterior surface of the humerus is the *triceps,* similar to the quadriceps extensor in the thigh and direct antagonist to the biceps and brachialis anticus muscles. It rises by a long head from below the glenoid fossa, by the external head from the upper third of the posterior surface of the humerus, and by the internal head from the middle and lower thirds of the posterior surface. It is inserted in the olecranon process of the ulna and serves to extend the forearm and arm.

The humerus is more often *fractured* by muscular action than any other bone. Usually the fracture occurs in the lower half of the bone and sometimes the musculo-spiral nerve is involved. There is a great tendency to non-union, probably due to interposition of soft parts. Sometimes the break is across and down between the condyles, *T-fracture*. Involvement of the elbow-joint is more serious than fracture of the humerus alone. *Sarcoma* of the humerus does occur and may require the removal of the clavicle and scapula as well as of the arm bone itself. In *amputation* of the humerus in children a long skin flap is left to allow for growth of the bone, as it is liable to grow again.

The Ulna.—In the forearm there are two bones, the ulna and the radius, of which the former is the longer. The ulna is on the inner side of the forearm and its upper end forms the greater part of the articulation with the humerus, as most of the articulation at the wrist is formed by the radius and the inter-articular fibro-cartilage. The head of the ulna is at the lower extremity of the bone and articulates on the outer side with the radius and below with the triangular fibro-cartilage. From its inner side projects the *styloid process*. The olecranon process forms the upper extremity and presents anteriorly an articular surface, the *greater sigmoid cavity*, for articulation with the trochlea of the humerus, where it fits into the olecranon fossa during extension. The same articulating surface also covers the *coronoid process*, a smaller projection below and in front of the olecranon, which fits into the coronoid fossa during flexion. Continuous with the greater sigmoid cavity on the outer side is the *lesser sigmoid cavity* for articulation with the head of the radius. Under the *triceps tendon*, which is inserted into the olecranon, is a *bursa* or sac of synovial membrane, such as occurs in parts where much force is brought to bear.

The Radius.—The radius, or spoke of the wheel, is on the outer side of the forearm and gets its name from the way it turns upon the ulna in pronation. The *shaft* is larger below than above and is slightly curved longitudinally for greater strength. The upper extremity or *head* is small and has a slightly concave upper surface for articulation with the radial head of the humerus. It articulates by its sides with the lesser sigmoid cavity and is bound to the ulna by the *orbicular ligament*, which runs over a smooth articular surface. Below the head is the constriction of the *neck* with the tuberosity for the biceps tendon to the inner side below. The lower extremity is large and forms the chief part of the wrist-joint, articulating with the semilunar and scaphoid bones of the wrist. From the lower extremity the strong conical *styloid process* projects externally.

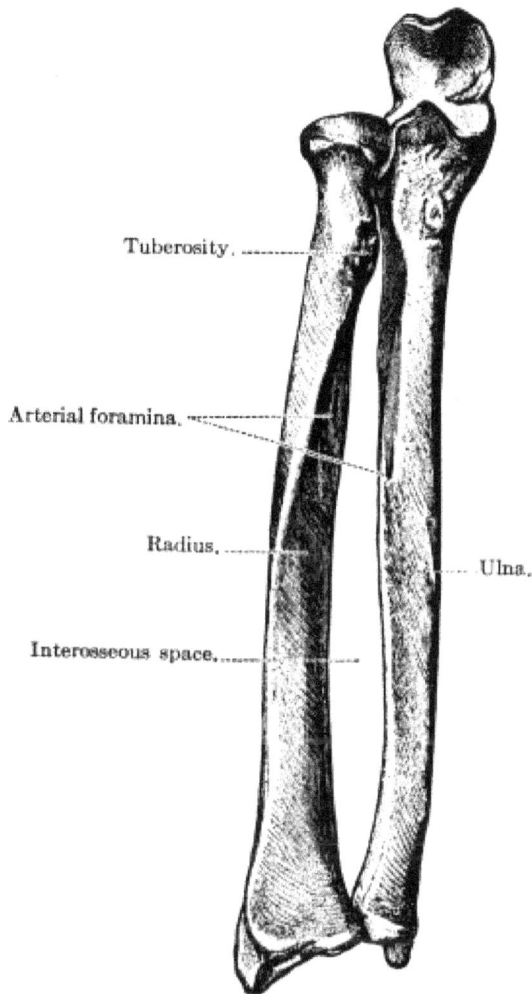

Fig. 72.—Bones of the right forearm
in a position of supination. (Toldt.)

In the living the *olecranon process* of the ulna is always felt at the elbow and the posterior border of the ulna forms the prominent ridge down the forearm, leading to the styloid process. The head of the radius is felt just below the external condyle and often makes a dimple in the muscles of the forearm. The rest of its upper half is concealed with muscles, but the lower half is easily felt as there are only tendons over it. The styloid process is felt externally. Normally that of the radius is a little lower than that of the ulna, so that in cases of fracture their relative position is of considerable importance as showing the amount of deformity.

The two forearm bones are more frequently *broken* together than separately and generally by direct violence, the lower fragment being usually drawn up by the action of the flexor and extensor muscles and producing a swelling on the palmar surface of the forearm. Indirect violence usually causes fracture of the radius only. In both cases, but especially in fracture of both bones, there is a tendency for membrane to get between the fragments, so the arm is put up in splints with the hand midway between pronation and supination in order to separate the bones as far as possible. Care must be taken not to have the bandage too tight or gangrene of the fingers may result. In most fractures of the arm it is put up bent, but in *fracture of the olecranon* it is put up fully extended, as the fragment is sure otherwise to be displaced by the pull of the triceps. In fact, the olecranon is sometimes fractured by the muscular force of the triceps, though usually its fracture, which is frequent, is due to direct violence. The ulna is also often fractured in the middle by direct violence or the styloid process may be broken. Fracture of the neck or shaft of the radius is very common, the most important arm fracture being that of the lower end of the radius or *Colles' fracture*. This and the corresponding fracture in the leg, *Pott's fracture*, are two of the commonest fractures. In *dislocation of the wrist* the normal relation of the two styloid processes remains unchanged, but in Colles' fracture the lower fragment often projects on the back of the hand, making a typical deformity called the *silver fork deformity*.

The bones of the wrist and hand had best be described before the forearm muscles are taken up, as the muscles of the forearm are distributed largely to the fingers.

The Wrist.—The wrist or _carpus_ is made up of eight bones arranged in two rows of four each. In the first row are the _scaphoid and semilunar bones_, on the outer side, articulating with the radius, the _cuneiform_ articulating with the fibro-cartilage of the wrist-joint, and the _pisiform_. In the second row, in corresponding positions, are the _trapezium, trapezoid, os magnum_, and *unciform*. The eminence felt on the radial side of the wrist is the protuberance of the scaphoid, while the pisiform is generally felt on the ulnar side.

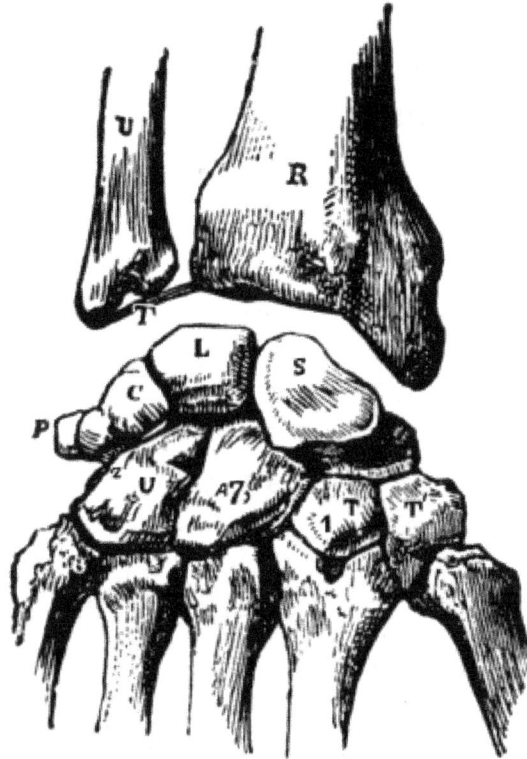

FIG. 73.—Right carpal bones, dorsal surface. *T*, trapezium; *T´*, trapezoid; *7*, os magnum; *U*, unciform; *S*, scaphoid; *L*, semilunar; *C*, cuneiform; *P*, pisiform.

The Hand.—The hand contains nineteen bones, five *metacarpal bones*, one for each finger and the thumb, whose bases articulate with the lower row of wrist bones, and fourteen *phalanges*, three for each finger and two for the thumb, of which the first row articulate with the metacarpal bones. They are all long bones and are slightly concave anteriorly. When the hand is flexed it is the heads of the metacarpal bones, not the bases of the phalanges, that are so prominent, the head of the third metacarpal being most prominent.

The metacarpals are seldom *fractured*, though bad fractures occasionally occur. In comminuted fracture nothing can be done but remove the bone. If the periosteum is left the bone will grow again. Two diseases sometimes affect the metacarpals and the phalanges, *tuberculosis* and *syphilis*. Both cause swelling of the bones.

Muscles of the Forearm.—The chief groups of muscles on the forearm are the *flexors* and *pronators* on the anterior surface and the *extensors* and *supinators* on the posterior surface. In general the flexors and pronators take their origin from on or around the internal condyle, while the extensors and supinators arise on or around the external condyle. Where not otherwise stated it will be understood that such is their origin. In a general way they may by grouped as follows:

ANTERIOR SURFACE.

Flexors of wrist	flexor carpi radialis flexor carpi ulnaris palmaris longus
Flexors of fingers	flexor sublimis digitorum flexor profundus digitorum
Flexor of thumb	flexor longus pollicis
Pronators of hand	pronator radii teres pronator quadratus

POSTERIOR SURFACE.

Extensor of forearm	anconeus
Extensors of wrist	extensor carpi radialis longior extensor carpi radialis brevior extensor carpi ulnaris
Extensors of thumb	extensor ossis metacarpi pollicis extensor primi internodii pollicis extensor secundi internodii pollicis
Extensor of fingers	extensor communis digitorum
Extensor of index finger	extensor indicis
Extensor of little finger	extensor minimi digiti
Supinators of hand	supinator longus supinator brevis

Of the *flexors of the wrist* the *flexor carpi radialis* is inserted into the base of the index and usually of the third metacarpal bone, the *flexor carpi ulnaris* into the fifth metacarpal, the pisiform and the unciform bones, while the *palmaris longus* goes to the anterior annular ligament of the wrist and the palmar fascia of the hand. The *flexor sublimis digitorum* is inserted by four tendons into the second phalanges of the fingers, while the *flexor profundus digitorum* arises from the upper part of the ulna and is inserted into the last phalanges of the fingers. The *flexor of the thumb* arises from the middle of the radius and is inserted into the last phalanx of the thumb. Which joint is flexed by a muscle depends upon the origin and insertion of the muscle, all those included between being affected. Thus, the *flexor sublimis digitorum*, which has its origin in part at least from the condyle and is inserted in the second phalanges of the fingers, flexes the forearm, wrist, and all the finger-joints but the last, while the *flexor profundus digitorum*, arising from the ulna, though it flexes the wrist and fingers, has no power of flexing the forearm.

The *pronator radii teres*, besides arising from the supra-condylar ridge, rises from the coronoid process of the ulna. It is inserted into the middle of the outer surface of the radius and serves to pronate the forearm. The other pronator, the *pronator quadratus*, is a small quadrilateral muscle extending transversely across the radius and ulna just above their carpal extremities. It rises from the anterior surface of the ulna and is inserted into the anterior external border of the radius.

On the back of the forearm the *anconeus* serves to extend the forearm only, being inserted into the upper part of the posterior surface of the ulna. The *extensors of the wrist* are inserted into the bases of the various metacarpal bones and have some power to extend the forearm as well as the wrist. The *extensors of the thumb*, as their names imply, go one to the metacarpal bone and one to each of the phalanges, the longest one extending the whole thumb, the others only a part. They rise from the ulna and radius, not the condyle. The *extensor communis digitorum* goes to all the phalanges of all the fingers, the *extensor minimi digiti* to those of the little finger only, and the *extensor indicis* to those of the index finger, the last two arising short of the condyle.

Of the *supinators* the longer one is inserted into the styloid process of the radius, while the shorter one, the *supinator brevis*, is inserted into the upper part of the same bone, both thus serving to turn the radius on the ulna.

Where the *tendons* of the various muscles pass over the wrist, both front and back, they are covered with a synovial sheath and are held down by a broad ligament, which some of them perforate, the annular ligament. The strong fibrous band of the *anterior annular ligament* arches over the carpal bones in front. Beneath it pass the median nerve and the tendons of the flexors of the fingers and thumb. The *posterior annular ligament* is of less importance.

The deep *palmar fascia* forms a sheath for the muscles of the hand. In carpenters there sometimes occurs *Dupuytren's contraction* of the palmar fascia, which draws the fingers up. As operation is not always successful, it is quite a serious matter.

The *muscles of the hand* itself include various abductor, adductor, and short flexor muscles of the thumb and little finger. There also extend between the metacarpal bones the *lumbricales*, four small muscles that aid the deep flexor muscles; likewise seven *interossei*, of which four are dorsal and three palmar. The *dorsal interossei* arise by two heads from the adjacent sides of the metacarpal bones and are inserted into the bases of the first phalanges, thus abducting the fingers; while the *palmar interossei*, arising from the palmar surface of the second, fourth, and fifth metacarpals, are inserted into the three corresponding first phalanges and adduct the fingers toward an imaginary line drawn through the middle finger.

Joints of the Upper Extremity.—The joints of the upper extremity, with the exception of the wrist-joint, are the most freely movable of any in the body, probably because the hand has the finest work to do and a greater number of motions are required. Even the wrist has much greater freedom of motion than the corresponding joint in the lower extremity.

The *shoulder-joint* is rather a deep joint, to allow of the varied motion required, and has a capsular ligament from the margin of the glenoid fossa above to the neck of the humerus below. The *elbow*, which is a hinge joint, has an anterior and a posterior ligament and two lateral ligaments, as is practically the case in all such joints. The *wrist* has several ligaments which, taken together, are capsular in nature.

Blood Supply of the Upper Extremity.—The blood supply of the upper extremity comes through the *subclavian artery*, which, on the right, springs from the innominate artery and on the left from the aortic arch. It remains one trunk as far as the elbow, though different names have been given to different parts. Thus, as it passes over the lower border of the first rib, it becomes the *axillary*, and at the lower border of the axilla, where it starts down the arm, the *brachial*. At the elbow it divides into the *ulnar* and *radial arteries*.

In its upper part the *brachial artery* lies internal to the humerus but below it is in front of the bone. The *radial* runs in a line from the middle of the elbow anteriorly to the inner side of the styloid process of the radius and is much exposed to injury in the lower third of its course, as when the hand is thrust through glass. On it at the wrist the pulse is counted. It is much smaller than the ulnar and winds around the outer side of the thumb to the palm, where, with the deep branch from the ulnar, it forms the *deep palmar arch*. The *ulnar artery* passes obliquely inward to the middle of the forearm and thence along its ulnar border to the palm of the hand, where it divides into the deep branch and the *superficial palmar arch* which supplies the four *digital arteries*.

From the *axillary artery* branches go to the chest wall and shoulder, the most important being the two *circumflex arteries* to the deltoid. The brachial has only two branches of any importance, the *superior* and *inferior profunda*, both on the upper arm, of course.

In case of *hemorrhage* compression can frequently be applied with the fingers where the subclavian crosses the rib or in the axilla, where the artery can be pressed up against the humerus.

Nerves.—The nerve supply of the shoulder comes chiefly from the anterior and posterior *thoracic*, the *suprascapular*, and the *circumflex*, these last going to the deltoid. The biceps is supplied by the *musculo-cutaneous*, the triceps by the *musculo-spiral*, and the brachialis anticus by both. Most of the flexor and pronator muscles are supplied by the *median*, while the *posterior interosseous* and the *musculo-spiral nerves* go to the extensors and supinators. The *ulnar nerve* supplies the hand largely.

CHAPTER XIII.
THE LOWER EXTREMITIES.

The lower extremities resemble the upper very closely in the arrangement of the bones, muscles, arteries, and nerves, though modifications occur, due to the difference in function of the lower limbs. There is one long bone in the upper part or thigh, the femur, and two in the lower part or leg, the tibia and fibula, while over the knee-joint is the patella or knee-cap. The ankle has seven bones and the foot nineteen like the hand.

The Femur.—The femur is the longest bone in the body, being about one-fourth the height of the person. It inclines toward its fellow at the knee in order to bring the knee-joints near the center of gravity in walking, the amount of inclination varying with the width of the hips and the height of the person. On account of the greater width of hip the tendency to knock-knee is greater in women than in men.

The *shaft* of the femur is enlarged at the extremities and is slightly curved forward, the concavity being strengthened at the back by a longitudinal ridge, the *linea aspera*, along part of which the gluteus maximus muscle is attached. The *head*, which is covered with cartilage, except for an oval depression for the attachment of the ligamentum teres, one of the ligaments of the hip-joint, and which articulates with the hollow of the acetabulum in the os innominatum, projects considerably upward, inward, and forward from the shaft, the *neck* varying much in length and angle. It is generally more horizontal in women than in men and in rickets the great weight on the softened bone tends to press the head down, causing the deformity known as "*coxa vera*", in which the neck is almost horizontal. Extending upward, outward, and backward from the shaft at the base of the neck, about three-quarters of an inch lower than the head and about on a level with the acetabulum and the spine of the os pubis, is the *greater trochanter*. This large, irregular prominence and the smaller one of the *lesser trochanter*, which is at the lower part of the base of the neck posteriorly, are for the attachment of muscles and to assist in rotating the bone. The lower extremity of the femur is larger than the upper and is flat from before backward. Between its two large eminences, the external and internal *condyles*, is a smooth depression in front, the *trochlear surface*, for articulation with the patella. The external condyle is more prominent in front, the internal inferiorly, the latter being the longer of the two by about half an inch. The *epiphysis* at the lower end of the femur is the only one in which ossification has begun at birth. Therefore, if ossification is found there, the child is known to have arrived at full term.

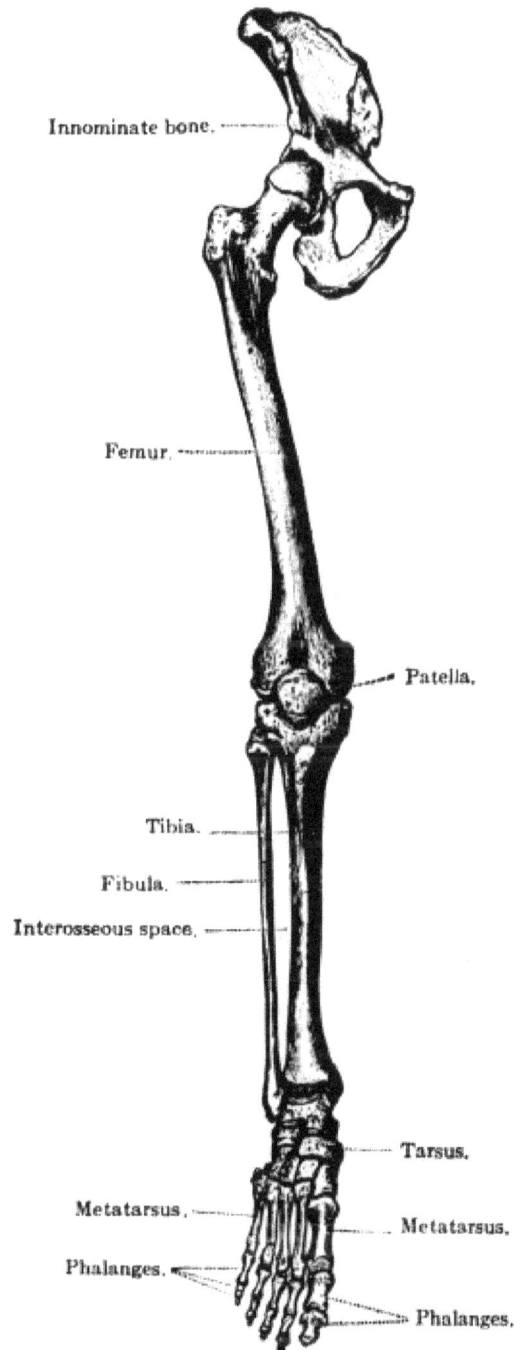

Innominate bone.

Femur.

Patella.

Tibia.

Fibula.

Interosseous space.

Tarsus.

Metatarsus.

Metatarsus.

Phalanges.

Phalanges.

So many large muscles are attached to the femur that the shaft cannot be detected in the living unless the person is very thin and poorly developed. The outer surface of the greater trochanter, however, and the condyles can be felt.

A string stretched from the anterior superior spine of the ilium to the tuberosity of the ischium passes in the middle just over the upper edge of the greater trochanter. The line thus drawn is known as *Nélaton's line* and is of considerable importance in many conditions of the hip. Thus, if the hip is dislocated, the trochanter will be thrown above *Nélaton's line,* and in osteomalacia the pelvis sinks and the trochanter is again above the line.

Thigh Muscles.—Of the thigh muscles only a few need be mentioned. One large muscle is the *psoas magnus*, which has its origin on the front of the last dorsal and all the lumbar vertebræ, passes down across the brim of the pelvis and under Poupart's ligament, gradually diminishing in size, and terminates in a tendon that is inserted into the lesser trochanter. It serves to flex the thigh on the pelvis and to rotate it outward. The *psoas parvus* rises from the last dorsal and the first lumbar vertebræ and does not go out of the pelvis.

The *sartorius or tailor muscle* is flat and ribbon-like and is the longest muscle in the body. It rises from the anterior superior spine of the ilium and is inserted into the upper inner surface of the shaft of the tibia. By it the legs are crossed. It also forms the outer side of an important landmark, *Scarpa's triangle*, whose base is formed by Poupart's ligament and the inner side by the *adductor magnus muscle*, which passes from the ramus of the os pubis and the tuberosity of the ischium to the linea aspera. The *femoral artery* bisects the triangle and runs into its apex.

The bulk of the anterior portion of the thigh is formed by the *quadriceps extensor,* which is really made up of four muscles, the *rectus femoris,* whose origin is on the anterior inferior iliac spine and above the acetabulum; the *vastus externus,* which comes from the greater trochanter and the upper linea aspera; and the *vastus internus* and *crureus,* which rise from the neck of the femur and the linea aspera. It is inserted into the tubercle of the tibia by the *ligamentum patellæ*, in which the patella lies. Its action is to extend the leg.

At the back and forming the *buttocks* are the three glutei muscles, the *gluteus maximus, medius, and minimus*. All these rise from the outer side of the ilium and have their insertion on or about the great trochanter. They serve to hold the trunk erect and to extend, abduct, and rotate the thigh.

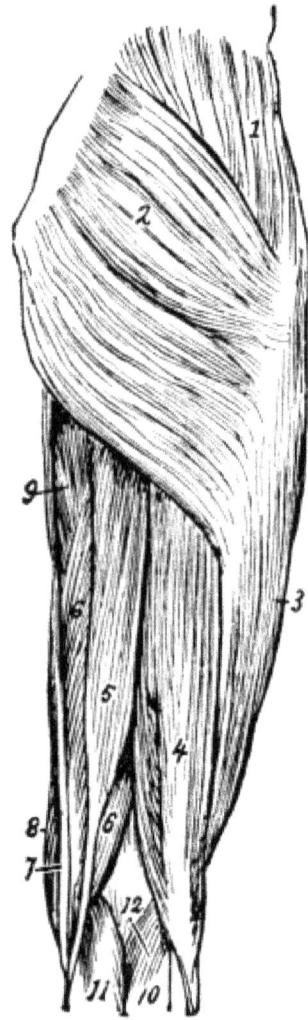

Fig. 75.

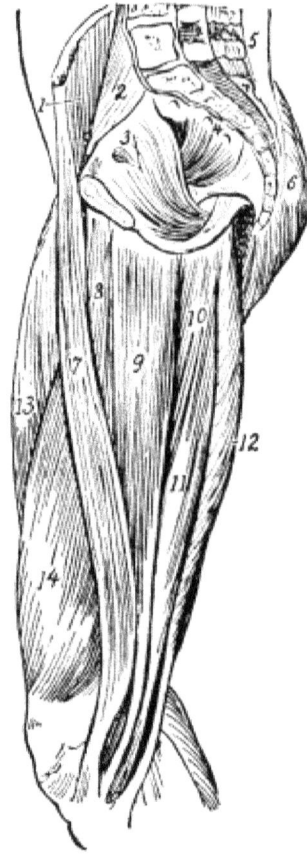

Fig. 76.

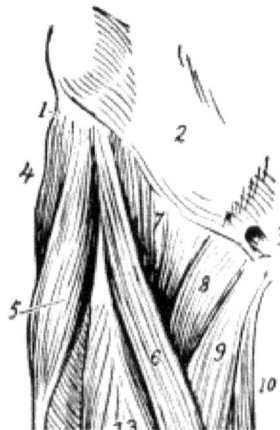

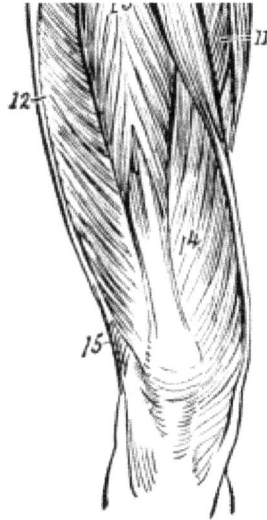

FIG. 77.

FIG. 75.—Superficial muscles of hip and thigh (from behind): 1, Gluteus medius; 2, gluteus maximus; 3, vastus externus; 4, biceps flexor cruris; 5, semitendinosus; 6, semimembranosus; 7, gracilis; 8, sartorius; 9, adductor magnus; 10, 11, gastrocnemius; 12, origin of plantaris. (Dorland's Dictionary.)

FIG. 76.—Muscles of the inner side of thigh and interior of pelvis: 1, Iliacus; 2, psoas magnus; 3, obturator internus; 4, pyriformis; 5, erector spinæ; 6, gluteus maximus; 7, sartorius; 8, adductor longus; 9, gracilis; 10, adductor magnus; 11, semimembranosus; 12, semitendinosus; 13, rectus femoris; 14, vastus internus. (Dorland's Dictionary.)

FIG. 77.—Superficial muscles of front of thigh: 1, Insertion of external oblique into iliac crest; 2, aponeurosis of external oblique; 3, external abdominal ring; 4, gluteus medius; 5, tensor vaginæ formoris; 6, sartorius; 7, iliopsoas; 8, pectineus; 9, adductor longus; 10, gracilis; 11, adductor magnus; 12, vastus externus; 13, rectus femoris; 14, vastus internus; 15, biceps flexor cruris. (Dorland's Dictionary.)

Lower down and forming the back of the thigh are the biceps and the semitendinosus and semimembranosus muscles. The *biceps* rises by two heads from the tuberosity of the ischium and the linea aspera and is inserted into the head of the fibula. It is on the outer side of the thigh and its tendon, which embraces the external lateral ligament of the knee-joint, forms the *outer hamstring.* On the inner side are the *semitendinosus* and the *semimembranosus muscles.* These rise from the tuberosity of the ischium and are inserted, the one into the upper inner surface of the shaft of the tibia and the other into the internal tuberosity of the tibia. Their tendons form the *inner hamstring*. Like the biceps they serve to extend the thigh and flex the leg on the thigh, but where the biceps rotates the leg out they, being attached to the inner side of the leg bones, rotate it in.

__The patella__, or *small pan*, is a flat, somewhat triangular bone developed in the quadriceps extensor tendon. Four *muscles* are attached to it as well as the ligamentum patellæ, which holds it to the tibia and gives increased leverage by making the quadriceps extensor work at a greater angle. It articulates with the condyles and serves to protect the joint. One bursa, the *prepatella bursa*, separates it from the skin and another, surrounded by adipose tissue, from the head of the tibia. The external surface can be seen and felt on the front of the knee and the bone can be moved from side to side when the leg is straight.

Joints of the Lower Extremity.—The *hip-joint* is a ball-and-socket joint but is not so freely movable as the shoulder-joint, the head of the femur being held in the acetabulum by many strong ligaments, of which the most important is the *capsular* ligament.

The *knee-joint* is largely a hinge joint, but in some positions it has some rotation. It is formed by the condyles of the femur, the head of the tibia, and the patella, and has fourteen ligaments, including the *ligamentum patellæ* and the *crucial ligaments*. Its synovial sac is the largest found in any joint. Two *semilunar cartilages*, placed on the head of the tibia, serve to deepen the socket for the condyles, changing somewhat in shape and thickness as the joint moves. The interval between the thigh and the leg bones can be felt at the knee. When the leg is extended the juncture of the bones is slightly above the patella, while in flexion a knife passed below the apex of the patella will pass into the joint.

Congenital dislocation of the hip occurs. *Separation of the epiphysis of the femur* may occur and sometimes the neck, rarely the lower part of the shaft, is *fractured*. Either condyle may be fractured off or there may be a T-fracture, in which case the popliteal artery may be injured. In *dislocation* the head may be behind or in front of the acetabulum. *Impacted hip*, where the neck of the femur has, in a fall, been driven into the head, is common in old people. Sometimes, especially in young children, the bone is infected, *osteomyelitis*. *Sarcoma* occurs. Most tubercular disease of the hip originates at the upper extremity of the femur, *tuberculosis* generally starting in the head and then attacking the capsule and the soft parts of the joint. If neglected, shortening of the leg may result, in which case the bone has to be broken and set at an angle in order to enable the child to walk.

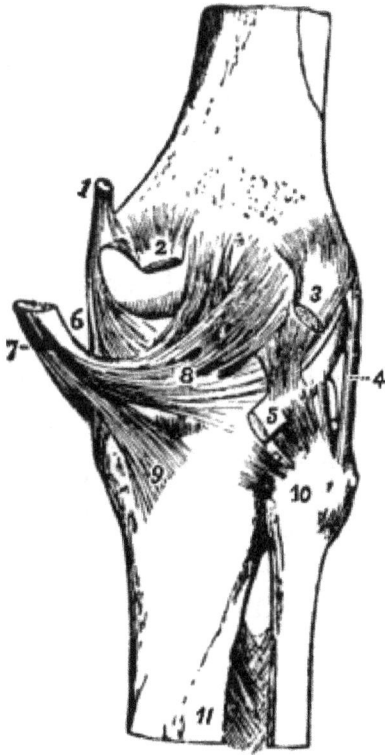

Fig. 78.

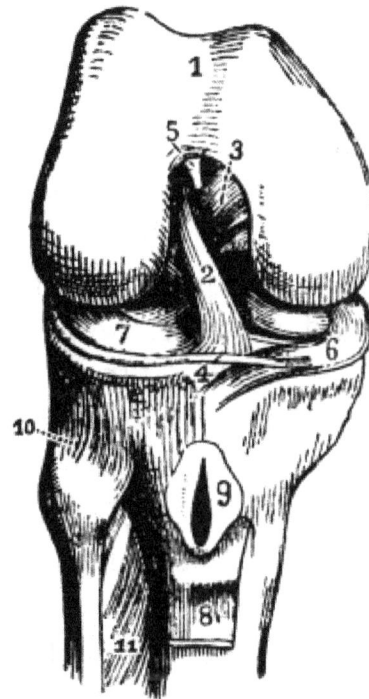

Fig. 79.

Fig. 78.—Right knee-joint, posterior view. (Leidy.)

Fig. 79.—Right knee-joint, showing internal ligaments: 2, anterior crucial ligament; 3, posterior crucial ligament; 4, transverse ligament; 6, 7, semilunar fibro-cartilages. (Leidy.)

Occasionally a bit of cartilage gets broken off in the knee-joint and wedged between the bones, so that the joint cannot be straightened. This is *dislocation of the semilunar cartilage* and necessitates an operation for removal of the piece. The cartilage will eventually be replaced by fibrous tissue and in a few months the leg will be all right. *Dislocation of the knee* is rare, though it may occur in any direction. Often the bursæ of the joint are irritated, as by kneeling to scrub floors, and *bursitis* or *housemaid's knee* results. *Fracture of the patella* may be caused by muscular traction or by direct violence, and is generally repaired by making an incision and sewing the parts of the bone together. *Tumor albus* or *white swelling* is *tuberculosis of the knee* and is fairly common in children. *Specific knee* means *syphilis of the knee* and generally occurs in both knees.

The Tibia.—The tibia or *shin bone* is next longest to the femur and is on the inner side of the leg, corresponding to the ulna in the arm. The *shaft* is prismoid and is more slender for the lower quarter, where fracture is consequently most frequent. The anterior border forms the *crest* or shin and can be felt for its upper two-thirds. The lower extremity, which is smaller than the upper, articulates with the astragalus bone of the ankle and with the fibula. Its *head* or upper extremity is expanded into two lateral *tuberositis* for articulation with the femur and for muscular attachment, both of which can easily be felt just below the bend of the knee. Their upper surfaces are smooth and concave, with a vertical bifid spine in the middle and a prominent tubercle for the attachment of the semilunar cartilages on either side. On the anterior surface of the head, below, is a rough eminence or tubercle, which also can be felt. The lower part of this is for the attachment of the ligamentum patellæ, while the upper part, which is smoother, is for the bursa that is placed under the tendon to prevent friction. On the back of the outer tuberosity is a facet for the head of the fibula. At the lower end there projects downward on the inner side, overhanging the arch of the foot, the *internal malleolus*, the prominent part of the ankle. It is on a higher level and somewhat farther forward than the *external malleolus*.

The Fibula.—The fibula is the most slender of all the bones in proportion to its length and is on the outer side of the leg. Its *head* is small and placed toward the back of the tibia below the knee-joint, from which it is excluded. The head articulates with the external tuberosity and has extending upward from it the styloid process. To it is attached the *biceps tendon* or *outer hamstring*. At the lower extremity of the shaft is the *external malleolus*, which articulates with the astragalus and forms the outer ankle. The only parts of the fibula that can be felt, besides the malleolus, which is very prominent, are the head and the lower external surface of the shaft.

In *fracture of the leg* both bones are usually broken, though either may be broken separately. *Pott's fracture* is fracture of the lower fibula, and may be caused by stamping hard when stepping on to the sidewalk. In rickets the tibia becomes bowed outward and forward, causing *bow leg,* a condition which in very young children may be rectified by manipulation. Later on braces are needed and after five years the bones have to be broken and set straight.

The Ankle.—The ankle or *tarsus* has but seven bones where the wrist has eight. They are the *os calcis* or *heel bone,* which is the largest and strongest and forms the tuberosity of the heel; the *astragalus,* which is next largest and helps to form the ankle-joint; the *cuboid*; the *navicular* (*boat-like*) or *scaphoid*; and the internal, middle, and external *cuneiform bones.* The astragalus is above and partially in front of the os calcis, to which is attached the *tendo Achillis.* The *cuboid* is on the outer side of the foot, in front of the os calcis and behind the metatarsals. It is noticeable in *congenital club-foot,* in which condition the tarsal bones may be distorted in shape and misplaced. The *navicular* or *scaphoid* is on the inner side of the foot, between the astragalus and the three cuneiform bones.

The Foot.—There are five *metatarsal bones* in the foot, corresponding to the five metacarpals in the hand, and the toes have the same number of *phalanges* as the fingers, though they are shorter and stronger. The big toe corresponds to the thumb.

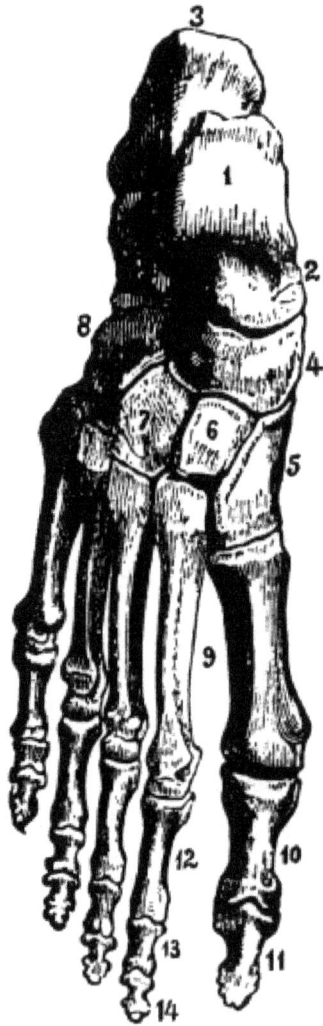

FIG. 80.—Bones of the right foot, dorsal surface: 1, Astragalus; 2, talus; 3, os calcis, 4, navicular; 5, internal cuneiform; 6, middle cuneiform; 7, external cuneiform; 8, cuboid; 9, metatarsus; 10-14, phalanges. (Leidy.)

Fracture of the os calcis and the astragalus are most commonly caused by a fall from a height, while the metatarsals and phalanges are generally broken by something heavy falling upon them. Because of their delicate structure, their distance from the heart, and the differences of temperature to which they are subjected, the tarsal bones are especially liable to become *tubercular*, amputation of the feet even becoming necessary at times. In diabetes there may be a perforating ulcer on the sole of the foot and the bone may become diseased.

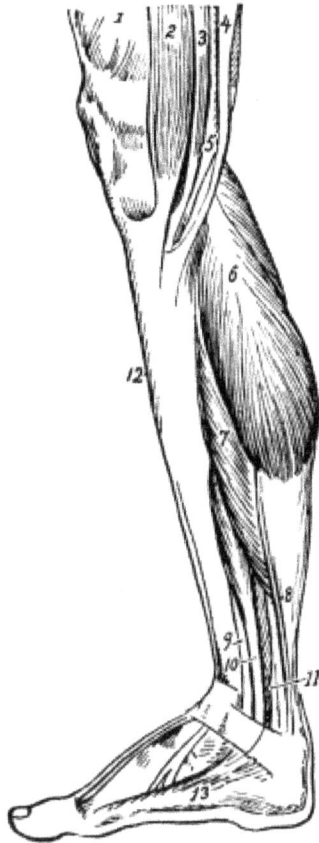

F<small>IG</small>. 81.

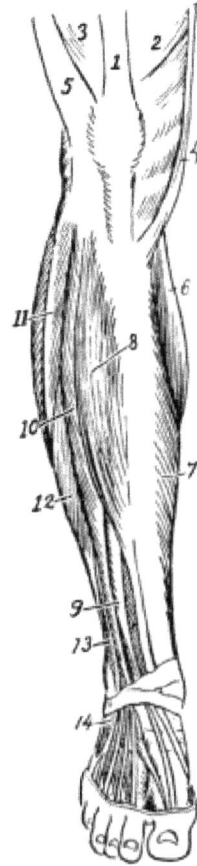

F<small>IG</small>. 82.

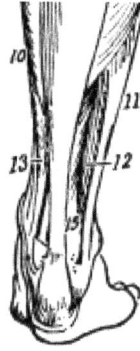

FIG. 83.

FIG. 81.—Superficial muscles of the leg from inner side: 1, Vastus internus; 2, sartorius; 3, gracilis; 4, semitendinosus; 5, semimembranosus; 6, inner head of gastrocnemius; 7, soleus; 8, tendon of plantaris; 9, tendon of tibialis posticus; 10, flexor longus digitorum; 11, flexor longus hallucis; 12, tibialis anticus; 13, abductor hallucis. (Dorland's Dictionary.)

FIG. 82.—Muscles of leg and foot (from before): 1, Tendon of rectus femoris; 2, vastus internus; 3, vastus externus; 4, sartorius; 5, iliotibial band; 6, inner head of gastrocnemius; 7, inner part of soleus; 8, tibialis anticus; 9, extensor proprius hallucis; 10, extensor longus digitorum; 11, peroneus longus; 12, peroneus brevis; 13, peroneus tertius; 14, origin of extensor brevis digitorum. (Dorland's Dictionary.)

FIG. 83.—Superficial muscles of leg (from behind): 1, Vastus externus; 2, biceps flexor cruris; 3, semitendinosus; 4, semimembranosus; 5, gracilis; 6, sartorius; 7, outer, and 8, inner, head of gastrocnemius; 9, plantaris; 10, soleus; 11, peroneus longus; 12, peroneus brevis; 13, flexor longus digitorum; 14, tibialis posticus; 15, lower fibers of flexor longus hallucis. (Dorland's Dictionary.)

Muscles of the Leg.—The greater part of the calf of the leg is formed by the _gastrocnemius_, a large bulging muscle, which rises from the condyles of the femur and is inserted along with the _soleus_, whose origin is on the back of the upper fibula, and the _plantaris_, which comes from the linea aspera, into the os calcis by a common tendon, the *tendo Achillis*, the largest and strongest tendon in the body. Its action is to extend the foot and to rotate it slightly inward. Other extensors of the foot, which also evert it, are the _peroneus longus_ and the _peroneus brevis_ at the upper and outer part of the leg, the former rising from the outer tuberosity of the tibia and the upper fibula and being inserted into the first metatarsal and the internal cuneiform, the latter arising from the lower fibula and being inserted into the fifth metatarsal. The foot is flexed, adducted, and rotated inward by means of the *tibialis anticus*, which rises from the outer tuberosity and the upper two-thirds of the outer surface of the tibia and is inserted into the internal cuneiform bone.

In the foot, and corresponding to the palmar fascia in the hand, is the *plantar fascia*, the densest of all fibrous membranes. There are also various annular ligaments, and the *foot muscles* are arranged similarly to those in the hand.

The Blood Supply of the Lower Extremity.—The blood supply of the lower extremity comes from the *external iliac artery*, a branch of the *common iliac*, which passes obliquely downward and outward along the border of the psoas muscle to Poupart's ligament, where it enters the thigh and becomes the *femoral artery*. Its only important branches are the *deep epigastric*, which goes up along the internal abdominal ring, and the *deep circumflex iliac*. As the *femoral artery* it passes down the inner side of the thigh to the internal condyle of the femur, being very superficial at Scarpa's triangle, where it can be compressed with the thumb to stop hemorrhage below. If a tourniquet is applied, it should be applied a little lower down. The first and most important branch of the femoral is the *profunda femoris*.

About two-thirds of the way to the knee the artery takes the name *popliteal*. It lies superficially in the popliteal space back of the knee, but above and below it is covered with muscles. Its branches supply the knee-joint and nearby muscles and are unimportant. At the lower border of the *popliteus muscle*, a small muscle at the knee, it divides into the *anterior and posterior tibial arteries*. The course of the former of these may be marked by a line from the inner side of the head of the fibula to midway between the malleoli at the front of the ankle, where it terminates in the *dorsalis pedis artery* for the back of the foot. By this last the pulse is sometimes taken and its pulsation is a guide in determining how high up to amputate in gangrene of the foot. The posterior tibial extends obliquely down the back of the leg to the heel, where it divides into the *internal and external plantar arteries* which go to the sole of the foot. Its most important branch is the *peroneal*.

Besides the *deep veins* accompanying the arteries there are the *superficial veins*, the *internal* or *long saphenous* on the inner side of the leg and thigh and the *external* or *short saphenous* on the middle of the leg posteriorly and emptying into the popliteal vein. Varicosity often occurs in these veins.

Nerves.—The nerves of the muscles about the hip are branches of the *lumbar nerve*. The *anterior crural* supplies the anterior part of the thigh, the *gluteal* the muscles of the same name, and the *great sciatic* the large muscles of the back of the thigh. Below the knee the *anterior tibial* goes to the tibialis anticus and the *internal popliteal* to the muscles of the calf, while the peroneus muscles are supplied by the *musculo-cutaneous*.

www.ingramcontent.com/pod-product-compliance
Lightning Source LLC
Chambersburg PA
CBHW081815200326
41597CB00023B/4252